Machining Process and Machines

Machining Processes and Machines

Fundamentals, Analysis, and Calculations

Zainul Huda

CRC Press
Taylor & Francis Group
Boca Raton London New York

CRC Press is an imprint of the
Taylor & Francis Group, an **informa** business

First edition published 2021

by CRC Press

6000 Broken Sound Parkway NW, Suite 300, Boca Raton, FL 33487-2742

and by CRC Press
2 Park Square, Milton Park, Abingdon, Oxon, OX14 4RN

© 2021 Taylor & Francis Group, LLC

CRC Press is an imprint of Taylor & Francis Group, LLC

Library of Congress Cataloging-in-Publication Data
Names: Huda, Zainul, author.
Title: Machining processes and machines : fundamentals, analysis, and calculations / Zainul Huda.
Description: First edition. | Boca Raton : CRC Press, 2020. | Includes bibliographical references and index.
Identifiers: LCCN 2020029951 (print) | LCCN 2020029952 (ebook) | ISBN 9780367532697 (hardback) | ISBN 9781003081203 (ebook)
Subjects: LCSH: Machining.
Classification: LCC TJ1185 .H793 2020 (print) | LCC TJ1185 (ebook) | DDC 671.3/5--dc23
LC record available at https://lccn.loc.gov/2020029951
LC ebook record available at https://lccn.loc.gov/2020029952

ISBN: 978-0-367-53269-7 (hbk)
ISBN: 978-1-003-08120-3 (ebk)

Typeset in Times
by SPi Global, India

Contents

Preface..xiii
Acknowledgements...xv
Author xvii

PART I *Fundamentals of Machining and Cutting Tools*

Chapter 1   Introduction 3

  1.1   What Is Machining? 3
  1.2   Industrial Applications of Machining 4
  1.3   Machining Processes: Classification and Types...........4
       1.3.1   Classification of Machining Processes...........4
       1.3.2   Conventional Machining Processes 4
       1.3.3   Abrasive Machining Processes 5
       1.3.4   Non-Traditional Machining...........................5
       1.3.5   Micro-Precision Machining and UPM...........6
  1.4   Roughing and Finishing in Machining 7
  1.5   Machinability and Machinable Materials 7
       1.5.1   Machinability 7
       1.5.2   Machinable Materials......................................8
  1.6   Industrial Importance of Calculations in Machining...............8
  Questions...9
  References 10

Chapter 2   Mechanics of Chip Formation...11

  2.1   Chip Formation in Machining.......................................11
  2.2   Orthogonal and Oblique Machining.............................11
       2.2.1   Orthogonal Machining 12
       2.2.2   Oblique Machining..12
  2.3   Engineering Analysis of Chip Formation.....................13
  2.4   The Effects of Rake Angle and Shear Plane Angle
        on Machining...15
  2.5   Merchant's Equation and Its Industrial Application...............16
  2.6   Types of Chips in Machining 16
       2.6.1   The Four Types of Chips...............................16
       2.6.2   Segmented or Discontinuous Chips 17
       2.6.3   Continuous Chips in Machining 17
       2.6.4   Other Types of Chips.....................................18
               2.6.4.1   Continuous Chips With BUE.....................18
               2.6.4.2   Serrated Chips 19

2.7   CALCULATIONS – Worked Examples on Mechanics
        of Chip Formation 19
        Questions and Problems 25
        References 26

Chapter 3   Forces and Power in Machining 27
        3.1   Importance of Cutting Forces and Power 27
        3.2   Forces on Work, Tool, and Chips during Machining 27
                3.2.1   Forces on Workpiece 27
                3.2.2   Forces on Tool/Chip 27
        3.3   Measuring the Cutting Force and the Thrust Force 28
        3.4   Mathematical Models for Forces in Machining 30
        3.5   Merchant's Force Circle 31
        3.6   Power and Energy in Machining 31
        3.7   Temperature in Machining 33
        3.8   Calculations – Worked Examples on Cutting Forces
                and Power 33
        Questions and Problems 41
        References 43

Chapter 4   Cutting Tools and Materials 45
        4.1   Cutting Tools and Materials: *Fundamentals* 45
                4.1.1   Cutting Tools and their Types 45
                4.1.2   Cutting Tool Materials 45
        4.2   Tool Wear and Failure 46
        4.3   Cutting Tool Life – *Taylor's Tool-Life Equation* 47
        4.4   Cutting Fluids 48
        4.5   CALCULATIONS – *Worked Examples on Cutting
                Tools and Materials* 49
        Questions and Problems 57
        References 59

Part II   Conventional Machining and Gear Manufacturing

Chapter 5   Turning Operations and Machines 63
        5.1   Turning Operations and their Industrial Applications 63
        5.2   Turning Related Operations 64
        5.3   Lathe – *Parts and Mechanism* 65
        5.4   Lathe Machine Tools 66
                5.4.1   Speed Lathes 66
                5.4.2   Engine Lathes 66
                5.4.3   Tool Room Lathe 66
                5.4.4   Turret Lathe 67

            5.4.5   Automatic Lathe..67
      5.5   Work-Holding Techniques in Lathe Practice....................68
      5.6   Engineering Analysis of Straight Turning..........................69
      5.7   Engineering Analysis of Taper Turning 72
      5.8   Engineering Analysis of Thread Turning...........................74
      5.9   CALCULATIONS – *Worked Examples on Turning*
            Operations and Machines 75
      Questions and Problems..87
      References 89

Chapter 6   Drilling Operations and Machines 91

      6.1   Drilling and Its Industrial Applications 91
      6.2   Drilling-Related Operations and Their Applications.............92
      6.3   Drill Bits...93
            6.3.1   Function, Materials, and Types of Drill Bit 93
            6.3.2   Twist Drill Bits 93
      6.4   Drilling Machines 94
            6.4.1   Features and Types of Drilling Machines.................94
            6.4.2   Upright Drill Press 94
            6.4.3   Radial Arm Drill Press 95
            6.4.4   Multiple-Spindle and Turret Drilling Machines 95
      6.5   Engineering Analysis of Drilling Operation 96
      6.6   CALCULATIONS – *Worked Examples on Turning*
            Operations..100
      Questions and Problems..105
      References 106

Chapter 7   Milling Operations and Machines 107

      7.1   Milling and Its Industrial Importance.................................107
      7.2   Forms of Milling – *Peripheral Milling and Face Milling* 107
      7.3   Methods of Milling – *Up Milling and Down Milling* 108
      7.4   Milling Machines – *Cutters and Types of Milling*
            Machines 108
            7.4.1   Milling Cutters...108
            7.4.2   Milling Machines and Their Types...........................109
                    7.4.2.1   Knee-and-Column-Type Milling
                              Machines..109
                    7.4.2.2   Universal Horizontal Milling Machines...110
                    7.4.2.3   Ram-Type Milling Machines....................111
      7.5   Milling Operations 111
            7.5.1   Operations on a Vertical-Spindle Knee-and-
                    Column-Type Milling Machine...............................111
            7.5.2   Operations on a Horizontal-Spindle Knee-and-
                    Column-Type Milling Machine................................111

7.6   Engineering Analysis of Milling.. 113
        7.6.1   Cutting Speed and Feed in Milling........................ 113
        7.6.2   Cutter's Rotational Speed, Table Feed Rate,
                  and MRR... 113
        7.6.3   Cutting Time in Milling... 114
        7.6.4   Cutting Power and Torque in Milling 115
7.7   Indexing in Milling 116
        7.7.1   Indexing and Indexing Head 116
        7.7.2   Methods of Indexing 117
                  7.7.2.1   Direct Indexing..................................... 117
                  7.7.2.2   Simple or Plain Indexing 117
                  7.7.2.3   Differential Indexing 118
7.8   CALCULATIONS – *Worked Examples on Milling*............. 119
Questions and Problems.. 129
References 131

Chapter 8   Shaping/Planing Operations and Machines.................................... 133

8.1   Shaping/Planing Operation and its Applications 133
8.2   Planing Machine Tool – Planer 134
        8.2.1   Planer – *Parts and their Functions* 134
        8.2.2   Working Principle of Planer.................................... 134
8.3   Shaping Machine Tool – *Shaper* 135
        8.3.1   Shaper – *Parts and their Functions* 135
        8.3.2   Quick Return Mechanism in a Shaper 136
8.4   Engineering Analysis of QRM in a Shaper........................... 136
8.5   Engineering Analysis of Shaping Operation 138
8.6   CALCULATIONS – *Worked Examples on
        Shaping/Shaper* 140
Questions and Problems.. 144
References 145

Chapter 9   Broaching and Broach Design.. 147

9.1   Broaching and its Applications 147
        9.1.1   Broaching – *An Introduction* 147
        9.1.2   Applications of Broaching...................................... 148
9.2   Broaching and its Design Analysis...................................... 148
        9.2.1   Broach and its Parts.. 148
        9.2.2   Broach Materials and Types.................................... 148
        9.2.3   Design analysis of Broach 149
9.3   Engineering Analysis of Broaching Operation 150
9.4   CALCULATIONS – *Worked Examples on Broaching* 151
Questions and Problems.. 154
References 155

Chapter 10 Gear Cutting/Manufacturing 157

    10.1   Gears – *Spur Gear Nomenclature*.. 157
    10.2   Gear Cutting Methods/Processes.. 158
    10.3   Gear Milling... 159
            10.3.1   Gear Milling Process 159
            10.3.2   Engineering Analysis of Gear Milling.................... 159
    10.4   Gear Broaching 161
    10.5   Gear Shaping 162
            10.5.1   Gear Shaping Process.. 162
            10.5.2   Engineering Analysis of Gear Shaping................... 163
    10.6   Gear Hobbing 164
            10.6.1   Gear Hobbing Process.. 164
            10.6.2   Hobbing Cutting Tool – *Hob*.................................... 164
            10.6.3   Hobbing Machine Tool.. 165
            10.6.4   Engineering Analysis of Gear Hobbing 166
                     10.6.4.1   Engineering Analysis of Single-Cut
                              Pass Gear Hobbing 166
                     10.6.4.2   Engineering Analysis of Double-Cut
                              Pass Gear Hobbing 166
    10.7   CALCULATIONS – *Worked Examples on Gear
          Manufacturing*... 168
            10.7.1   Worked Examples on Gear Milling 168
            10.7.2   Worked Examples on Gear Shaping........................ 170
            10.7.3   Worked Examples on Gear Hobbing........................ 171
    Questions and Problems 175
    References 176

PART III *Grinding/Abrasive Machining Processes*

Chapter 11 Grinding Operations and Machines.. 181

    11.1   Grinding Operation and Its Advantages.............................. 181
    11.2   Grinding Wheel and Its Parameters 182
            11.2.1   Grinding Wheel – *Grains Cutting Action*............... 182
            11.2.2   Grinding Wheel Parameters.................................... 183
    11.3   Surface Finish – *Achieving a Good Surface Finish
          in Grinding* 184
    11.4   Surface Roughness/Quality 185
            11.4.1   Surface Quality 185
            11.4.2   Surface Roughness – *Parameters and
                     Calculation*... 185
            11.4.3   Surface Roughness Measurement/Testing 186

11.5   Grinding (Machining) Machines 187
        11.5.1   Grinding Machines Features and Their Types......... 187
        11.5.2   Surface Grinding Machines 187
        11.5.3   Cylindrical Grinding Machines 187
        11.5.4   Centerless Grinding Machines............................... 188
11.6   Engineering Analyses of Grinding Operations 189
        11.6.1   Engineering Analysis of Surface Grinding............. 189
        11.6.2   Engineering Analysis of Cylindrical Grinding........ 191
11.7   CALCULATIONS – *Worked Examples on Grinding
        Operations* 192
Questions and Problems 198
References 200

Chapter 12 Abrasive Finishing Machining Operations...................................... 201
12.1   Abrasive Finishing Machining and their Applications 201
12.2   Honing and its Engineering Analysis.................................. 201
        12.2.1   Honing Operation and its Applications.................... 201
        12.2.2   Mathematical Modeling of Honing 203
12.3   Lapping – *Operation, Advantages, Applications, and
        Analysis* 203
        12.3.1   Lapping Operation 203
        12.3.2   Advantages and Applications of Lapping 204
        12.3.3   Mathematical Modeling of Lapping Operation 204
12.4   Superfinishing...205
12.5   Polishing 205
12.6   CALCULATIONS – *Worked Examples on Abrasive
        Finishing Operations*..206
Questions and Problems 210
References 211

Part IV   Advanced/Non-Traditional Machining

Chapter 13 Computer Numerically Controlled Machining................................215
13.1   What Is Computer Numerical Control Machining?............. 215
13.2   Advantages and Limitations of CNC Machines................... 215
13.3   Computer-Aided Design (CAD) in CNC Machining 217
        13.3.1   What Is CAD?... 217
        13.3.2   Role of CAD in CAM/CNC Machining................... 217
13.4   CNC Machine – *Working System* 217
        13.4.1   G-Codes and Work-Part Program 219
                  13.4.1.1   Code Words in a Part Program................ 219
                  13.4.1.2   Work-Part Program in CNC
                             Programming.. 219
        13.4.2   Machine Control Unit 220

13.4.3   Processing Equipment in CNC Machining 221
13.5   Zero Systems and Positioning Systems
in CNC Machining 221
13.5.1   Zero Systems in CNC Machining 221
13.5.2   Coordinate Positioning Systems............................ 222
13.5.3   Positioning System 222
13.6   Control Systems in CNC Machines................................ 223
13.6.1   Open-Loop Control CNC System 223
13.6.2   Closed-Loop Control CNC System 224
13.7   Mathematical Modeling of Open-Loop Control CNC
Machining.. 224
13.8   Mathematical Modeling of Closed-Loop CNC System 225
13.9   CALCULATIONS – *Worked Examples on
CNC Machining*.. 226
Questions and Problems 232
References 233

Chapter 14   Mechanical Energy-based Machining Processes 235

14.1   What are the Mechanical Energy-Based Machining
Processes?.. 235
14.2   UltraSonic Machining (USM)... 235
14.2.1   The USM Process and Its Advantages 235
14.2.2   Mathematical Modeling of USM 236
14.3   Abrasive Jet Machining (AJM) 237
14.3.1   AJM Process and Its Advantages............................ 237
14.3.2   AJM Process Parameters – *Achieving
Higher MRR/Better Surface Finish* 238
14.3.3   Mathematical Modeling of AJM............................ 239
14.4   Water Jet Machining 239
14.4.1   Water Jet Machining Process and Its Advantages ... 239
14.4.2   Mathematical Modeling of Water Jet
Machining 240
14.5   Abrasive Water Jet Machining (AWJM) 241
14.5.1   AWJM *Process and Its Advantages* 241
14.5.2   Mathematical Modeling of AWJM 242
14.6   CALCULATIONS – *Worked Examples on Mechanical-
Energy Machining* 242
Questions and Problems 248
References 249

Chapter 15   Thermal Energy-based Machining Processes................................... 251

15.1   What Are Thermal Energy Machining Processes?............... 251
15.2   Electric Discharge Machining (EDM) 251

15.2.1 The EDM Process and Its Applications 251
15.2.2 Mathematical Modeling of EDM 252
15.3 Electric Discharge Wire Cutting (*EDWC*) 252
15.3.1 The EDWC Process and Its Advantages 252
15.3.2 Mathematical Modeling of EDWC 253
15.4 Electron Beam Machining (EBM) 254
15.4.1 The EBM Process 254
15.4.2 Mathematical Modeling of EBM 254
15.5 Laser Beam Machining (LBM) 255
15.6 Plasma Arc Cutting (PAC) 255
15.6.1 The PAC Process and Its
        Advantages/Limitations 255
15.6.2 Mathematical Modeling for the Operating
        Cost in PAC 257
15.7 CALCULATIONS – *Worked Examples on Thermal
      Energy Machining* 257
Questions and Problems 261
References 262

Chapter 16 Electrochemical Machining and Chemical Machining
               Processes 263

16.1 Electrochemical Machining (ECM) Processes 263
16.2 Electrochemical Machining 263
16.2.1 The ECM Process and Its Advantages 263
16.2.2 Chemical Reactions in ECM Process 264
16.2.3 Mathematical Modeling of ECM 265
16.2.4 Machining Parameters in ECM to Achieve a
        Good Surface Finish 266
16.3 Electrochemical Deburring (ECD) 267
16.4 Electrochemical Grinding (ECG) 267
16.5 Chemical Machining 268
16.6 CALCULATIONS – *Worked Examples on
      Electrochemical Machining* 269
Questions and Problems 272
References 273

Answers 275
Index 279

Preface

Machining/Material Removal Processes is taught as a full Manufacturing-2 course in undergraduate Mechanical Engineering programs in almost all reputed universities. This book, therefore, covers fundamentals and engineering analysis of both conventional and advanced material removal processes, along with gear cutting/manufacturing technology and computer numerically controlled (CNC) machining.

This textbook provides a holistic understanding of machining processes and machines, and enables critical thinking through mathematical modeling and problem-solving. The salient features of this text include 200 calculations/worked examples, 70 MCQs (with answers), 120 diagrams/photographs, and around 200 equations on machining operations, including gear cutting and CNC machining. The book is equally useful to both engineering-degree students and production/design engineers practicing in manufacturing industry; the latter may specially benefit from highly practical stuff on gear cutting/manufacturing (Chapter 10).

The book is divided into four parts. *Part I* (Chapters 1–4) first introduces readers to machining, machinability, types of machining processes, and theory of metal cutting, followed by a discussion on cutting tools and materials. In this *Part*, a detailed analysis of mechanics of chip formation and cutting forces are presented. *Part II* (Chapters 5–10) deals with conventional machining operations, including gear cutting/manufacturing. Here, special emphasis is given to turning, drilling, milling, and broaching, including their engineering analyses and problem solving.

Part III (Chapters 11 and 12) first introduces readers to grinding wheels and their specifications, followed by discussions on grinding operations and abrasive finishing machining operations. Here, the techniques to achieve a good surface finish by controlling grinding parameters are established. In this *Part*, grinding operations are first described and mathematically modeled, and grinding machines' working is illustrated; then abrasive finishing machining operations (honing, lapping, super-finishing, etc.) are explained, including their engineering analysis.

Lastly, *Part IV* (Chapters 13–16) presents advanced machining technologies. Here, computer numerically controlled (CNC) machines are introduced, along with the mathematical modeling of CNC machining. The chapter on CNC machining covers both computer aided design (CAD) and computer aided manufacturing (CAM), along with their mathematical modeling and problem solving. This *Part* also deals with non-traditional machining processes; these include ultra-sonic machining (USM), water jet machining (WJM), abrasive water jet machining (AWJM), abrasive jet machining (AJM), electric discharge machining (EDM), electric discharge wire cutting (EDWC), electron beam machining (EBM), laser beam machining (LBM), electro-chemical machining (ECM), and chemical machining (ChM). Almost all the non-traditional material removal processes are mathematically modeled.

This book contains 16 chapters. Each chapter first introduces readers to the technological importance of the topic and provides basic concepts with diagrammatic illustrations, and then its engineering analysis/mathematical modeling, along with calculations are presented. An attempt is made to use SI units in each mathematical model throughout the text. There are altogether **200 worked examples/calculations, 120 diagrams/photographs, 70 MCQs (with answers)**, and around **200 mathematical models/equations.** In particular, the worked examples cover solutions to both simple and difficult-to-solve problems. In order to best benefit from this book, the student/reader is advised to first prepare a list of relevant formulae and then attempt to solve a problem (worked example) while hiding its solution; they must try to solve the problem on their own by analyzing the data. In case the reader is stuck during solving the problem, they may refer to its solution. *Question and Problems* are included at the end of each chapter. The end-of-chapter problems must also be solved by the student/reader following the verification of their respective answers; the latter are provided at the end of the book. Based on my long-term associations with academia and industry, I anticipate that this volume proves a useful literature for both engineering students and practicing engineers. In particular, engineers/machine shop managers may greatly benefit from the 200 worked examples/calculations.

Zainul Huda
Department of Mechanical Engineering
King Abdulaziz University, Saudi Arabia

Acknowledgements

Thanks to God who blessed me with the wisdom that enabled me to write this book. I appreciate my wife for her extended cooperation during the write-up of this book. I would like to acknowledge Syed Amir Iqbal, PhD, Professor and Dean, Mechanical & Manufacturing Engineering, NED University of Engineering & Technology, Pakistan, for providing the high resolution picture of CNC machines installed in the computer aided manufacturing (CAM) Lab of his Department of Industrial & Manufacturing Engineering for Chapter 13. I am also grateful to Sarosh Lodi, PhD, Professor and Vice Chancellor, NED University of Eng. & Tech., for granting permission to publish the picture of the CNC machines in this book. I am thankful to Adil Abbas Ashary, PhD, Senior Principal Engineer, Bloom Energy Corporation, San Jose, California, USA, for valuable discussion with him on abrasive jet machining (AJM). I am grateful to Mohammed Salah Abd-Elwahed, PhD, Assistant Professor, Mechanical Engineering Department, King Abdulaziz University (*KAU*), Jeddah, Saudi Arabia, for providing the surface roughness tester for conducting surface grinding experiments that resulted in the contribution to Chapter 11. I am indebted to Talha, Machine Shop Supervisor, Department of Mechanical Engineering, *KAU*, for his assistance in the experimental work on the measurement of cutting forces in turning for Chapter 3. I am also thankful to Mosab Muhammed, Senior Year Mechanical Engineering student, *KAU*, for providing pictures of the dynamometer and machine tools for Chapters 3 and 5, respectively. I also appreciate Faiz-ul Huda, an Alumni of Nilai University, Malaysia, for his assistance in the rechecking/review of calculations throughout the book's manuscript. I am indebted to Shoaib Ahmed Khan, Financial Analyst, *UniLever PLC*, Jeddah, Saudi Arabia, for his assistance in developing high-quality graphical plots for inclusion in Chapter 4.

Author

Besides this book, **Professor Zainul Huda** (ORCID ID: 0000-0002-3433-4995) is the author of 6 books, including *Manufacturing: Mathematical Models, Problems, and Solutions* (CRC, Press, 2018) and *Metallurgy for Physicists and Engineers Fundamentals, Applications, and Calculations* (CRC Press, 2020). Dr. Zainul Huda is a Professor of Manufacturing Technology at the Mechanical Engineering Department, King Abdulaziz University (KAU), Jeddah, Saudi Arabia. His teaching interests include Material Removal Processes, Manufacturing Technology, Metallurgy, and Mechanical Behavior of Materials. He (as a *PI*) has recently completed a funded research project on autonomous robots. He is also a *PI* in another SAR 35,000 research project on boilers' super-heater tubes material. He has been working as a full Professor in reputed world ranking universities (including University of Malaya, King Abdulaiz University, King Saud University, etc.) since February 2007. Prof. Huda has 40 years' professional experience in materials, manufacturing, and mechanical engineering. He is a Professional Engineer (PE) registered under Pakistan Engineering Council. He has worked as Plant Manager, Development Engineer, Metallurgist, and Graduate Engineer in various manufacturing companies, including Pakistan Steel Mills Corporation Ltd. Prof. Huda earned a PhD in Materials Technology from Brunel University, London, UK, in 1991. He is also a postgraduate in Manufacturing Engineering. He obtained B. Eng (Metallurgical Engineering) from the University of Karachi, Pakistan, in 1976.

Prof. Huda is the author/co-author of a patent and 130 publications, which include 7 books, 36 international peer-reviewed journal articles (27 ISI-indexed journal articles), and many conference proceedings in the fields of materials, manufacturing, and mechanical engineering, published by reputed publishers from the USA, Canada, the UK, Germany, France, Switzerland, Pakistan, Saudi Arabia, Malaysia, South Korea, and Singapore. He has been cited over 800 times in Google Scholar. His author's h-index is 13 (i10: 18). He (as a *PI*) has attracted nine (9) research grants, worth total of USD 0.16 million. He has also successfully completed 20+ industrial consultancy/R&D projects in the areas of failure analysis and manufacturing in Malaysia and Pakistan. He is the developer of Toyota Corolla cars' axle-hub's heat treatment procedure, first-ever implemented in Pakistan (through Indus Motor Co/ Transmission Engineering Industries Ltd, Karachi). Prof. Huda has delivered guest lectures in the UK, South Africa, Saudi Arabia, Pakistan, and Malaysia. He is a Member of prestigious societies, including Institution of Mechanical Engineers (*IMechE*), London, UK; Canadian Institute of Canadian Institute of Mining, Metallurgy and Petroleum (CIM), Canada; and Life-time Member of Pakistan Engineering Council. Prof. Zainul Huda is the world's Top 20 (#14) scholar in the field of *Materials and Manufacturing* (see www.scholar.google.com).

Part I

Fundamentals of Machining and Cutting Tools

1 Introduction

1.1 WHAT IS MACHINING?

Machining is a subtractive manufacturing process that involves material removal, usually in the form of chips, from a workpiece. In machining, material is removed from a workpiece by using either a cutting tool or an energy source. In traditional machining, material removal is accompanied by the formation of chips, which is accomplished by use of a cutting tool with cutting edge(s). On the other hand, non-traditional machining (NTM) processes are chip-less material removal processes that involve the use of energy for material cutting. The traditional machining operations include turning, drilling, milling, shaping/planing, broaching, grinding, and the like (see Section 1.3).

Machining may be considered as a system consisting of the workpiece, the cutting tool, and the equipment (machine tool). In machining, there exists a relative motion between the tool and the work; the primary motion is called *cutting speed*, whereas the secondary motion is called *feed*. In general, there are three fundamental cutting conditions: (a) cutting speed, (b) feed, and (c) depth of cut. *Cutting speed* is the greatest of the relative velocities of cutting tool or workpiece. For example, in turning machining operation, the surface speed of the workpiece is the *cutting speed* (*v*), usually expressed in m/min (see Figure 1.1). *Feed* (*f*) is the distance moved by the tool (or by the work) per revolution, usually expressed as mm/rev. *Depth of cut* (*d*) is the distance the cutting tool penetrates into the work.

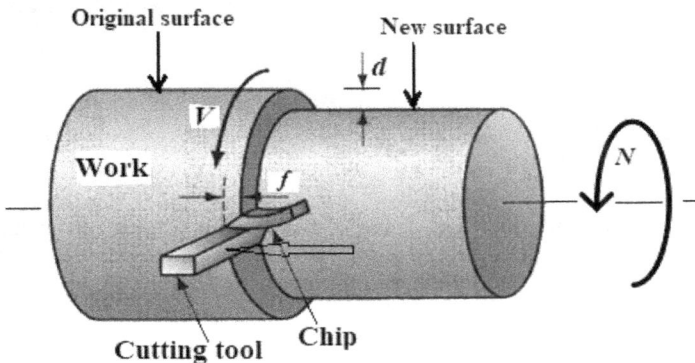

FIGURE 1.1   Turning operation (*d* = depth of cut; *f* = feed, *V* = surface speed, *N* = rotational speed)

1.2 INDUSTRIAL APPLICATIONS OF MACHINING

A wide range of engineering industries heavily rely on machining operations for manufacturing their products. In fact, machining is one of the eight basic manufacturing processes (metal casting, polymer/ceramic molding, metal forming, powder metallurgy, machining, welding/joining, heat treatment, and surface engineering). In general, cast, formed, and molded products require machining/material removal to achieve accurate dimensions and desired finish. The material removal machines are used for cutting metallic and non-metallic solids (e.g., metal, ceramics, wood, plastics, etc.). These machines find wide applications in diversified engineering industries. For example, the aerospace industry relies on machines with a high level of precision and accuracy to manufacture aircraft parts that fit the exact specifications of aeronautical design. Biomedical manufacturing companies produce life-saving devices that are used in hospitals and clinics across the world. The manufacturing of these life-saving devices and many other engineering components involves a great deal of machining operations. For example, gears and many machine elements can be manufactured by milling machining operations (see Chapters 7 and 10).

There are a number of advantages of machining processes which include (a) variety of work materials can be machined; (b) variety of part shapes and special geometric features are possible, such as screw threads and accurate round holes; (c) very straight edges and surfaces can be obtained; (d) good dimensional accuracy; (e) good surface finish; and (f) special geometric details can be created. In particular, the use of computer numerically controlled machines and other advanced machining technologies enables us to machine parts with high precision and productivity. However, there are also some limitations of machining/material removal processes; these disadvantages mainly include (a) wastefulness of materials in the form of chips, (b) time consumed in machining, and (c) the cost of machining.

1.3 MACHINING PROCESSES: CLASSIFICATION AND TYPES

1.3.1 CLASSIFICATION OF MACHINING PROCESSES

There exist a large number of material removal processes to fulfill the basic requirements of machining. All machining processes can be broadly classified into four categories: (a) conventional machining processes, (b) grinding and other abrasive machining processes, (c) NTM processes, and (d) micro-precision and ultra-precision machining (UPM) processes (see Figure 1.2). The four groups of machining processes are briefly discussed in the following sub-sections.

1.3.2 CONVENTIONAL MACHINING PROCESSES

Conventional machining processes involve the use of machine tools, such as lathes, milling machines, drill presses, shaper, or others, with a sharp cutting tool that removes material to achieve the desired geometry. The cutting tool may be either a single-point tool or a multiple-point cutting tool. Notable examples of conventional machining processes include *turning*, *drilling*, *milling*, *shaping*, *broaching*, and

Machining Processes

Conventional machining
- Turning
- Drilling
- Milling
- Shaping/Planing
- Broaching
- Sawing
etc.

Abrasive machining
— Grinding
  - Surface grinding
  - Cylinderical grinding
  - Centerless grinding
— Abrasive finish machining
  - Honing
  - Lapping
  - Superfinishing
  - Polishing

Nontraditional machining
—Mechanical energy mach. processes
—Thermal energy machining processes
— Electrochemical machining processes
— Chemical machining

Microprecision machining and UPM
— Microprecision machining
— Ultra-precision machining

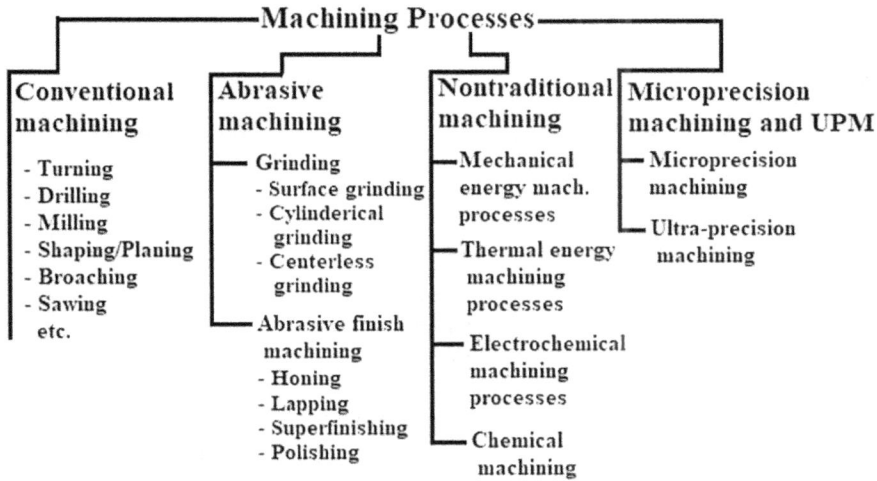

FIGURE 1.2 Classification of machining processes

the like. *Turning* involves the removal of material from a rotating cylindrical work-piece by the use of a single-point cutting tool; the latter has a feed motion (see Figure 1.1). *Milling* involves the use of a rotating cutter with multiple-point cutting edges; here, the cutting tool has the speed motion, and the work has feed motion. A detailed account of conventional machining processes is given in Part II (Chapters 5–10).

1.3.3 ABRASIVE MACHINING PROCESSES

An abrasive machining process involves the use of a grinding wheel, an abrasive stick, or an abrasive suspension to remove material from a workpiece. *Grinding* is a material removal operation by the action of hard and abrasive particles that are bonded usually in the form of a wheel (grinding wheel). As indicated in Figure 1.2, a grinding operation may be surface grinding, cylindrical grinding, or center-less grinding. The abrasive finishing machining operations include honing, lapping, superfinishing, and the like. Grinding and other abrasive machining processes are explained in detail in Part III (Chapters 11 and 12).

1.3.4 NON-TRADITIONAL MACHINING

Non-traditional machining (NTM) processes are chip-less material removal pro-cesses that involve use of energy for material cutting. The NTM processes may be classified into four groups: (a) mechanical energy-based machining processes, (b) thermal energy-based machining processes, (c) electro-chemical machining (ECM) processes, and (d) chemical machining process (see Figure 1.2). The mechanical energy machining involves the removal of material by mechanical erosion or abrasion; these processes include ultra-sonic machining (USM), water jet machining (WJM), abrasive WJM, and abrasive jet machining (AJM). Thermal energy machining relies on heat energy that causes removal of material by melting or vaporization; these processes include electric discharge machining, electric discharge wire cutting, electron beam machining, and laser beam machining. ECM processes are based on anodic metal dissolution in the presence of an electrolyte. The NTM processes are discussed in detail in Chapters 14–16.

1.3.5 MICRO-PRECISION MACHINING AND UPM

Micro-precision machining and UPM are playing a vital role in advanced manufacturing. *Micro-precision machining* is defined as "a manufacturing technology that involves the use of mechanical micro-tools with geometrically defined cutting edges in the subtractive fabrication of devices or features with at least some of their dimensions in the micrometer range (1–999 μm)" (Mativenga, 2014). Micro-precision machining processes enable us to produce micro-components, micro-features, and micro-structures.

Ultra-precision machining (*UPM*) is defined as "a machining process whose accuracy has been driven to its ultimate technological limits, irrespective of the nature of the process and the size of the workpiece (macro-, micro-, or nanoscopic)" (Brinksmeier, 2014). UPM processes allow us to achieve surface roughness down to nanometer scale; they are applied to manufacture optical components and similar components (Kang and Ahn, 2007; Schneider et al., 2019). For example, *diamond turning*, also called *single-point diamond turning*, is a *UPM* technology that involves the use of geometrically defined single-point diamond cutting tools for generating complex functional surfaces by turning operation (see Figure 1.3).

FIGURE 1.3   Machining of a complex geometry (conical shaped) work by diamond turning

1.4 ROUGHING AND FINISHING IN MACHINING

In conventional machining processes, it is often desirable to remove high stock from the bulk material (solid) as well as to achieve reasonably good surface quality. However, the achievement of both the objectives in a single pass is not possible. Thus, machining is usually carried out in two steps with varying cutting conditions (cutting speed, feed, and depth of cut). The two steps in machining are (a) roughing pass and (b) finishing pass. In the *roughing pass*, a bulk amount of material is quickly removed from the workpiece as per required feature. In this step, higher feed rate and depth of cut are employed so as to achieve a high material removal rate from the work. The roughing pass creates a shape close to desired geometry but leaves some machining allowance (material unremoved) for finish cutting. The roughing pass cannot provide good surface finish and close tolerance. This is why a *finishing pass* is carried out to improve surface finish, dimensional accuracy, and tolerance level; here, the feed rate and depth of cut are low. Thus, the material removal rate (MRR) is reduced in the finishing pass, but the surface quality is improved. Table 1.1 presents the main differences between roughing and finishing in conventional machining processes.

1.5 MACHINABILITY AND MACHINABLE MATERIALS

1.5.1 MACHINABILITY

Machinability refers to the ease of machining a material to obtain desired results at low cost. There are a number of quantitative measures of machinability; these measures include (a) tool life, (b) surface finish, and (c) other measures, such as cutting force, power, temperature, and chip formation. The tool life refers to the service time in minutes or seconds to a total failure of the cutting tool at certain cutting speed. The surface finish refers to the acceptable surface finish produced at standardized cutting speeds and feeds. A good machinability may mean one or more of the following: (a) minimum cutting forces, power, and temperature; (b) longer tool life (minimum tool wear); and (c) a good surface finish.

TABLE 1.1
Difference between Roughing and Finishing

Roughing in Machining	Finishing in Machining
It is performed prior to finishing pass	It is performed only after roughing pass
Its objective is to remove bulk excess material from workpiece in every pass	Its objective is to improve surface finish, dimensional accuracy, and tolerance
It involves higher feed rate and depth of cut	It involves higher cutting speed
It results in higher MRR	The MRR is low
The surface finish is poor	The surface finish is good
It cannot provide high dimensional accuracy and close tolerance	It can provide high dimensional accuracy and close tolerance
It permits the use of an old cutter	It requires the use of a sharp cutting tool

1.5.2 Machinable Materials

In order to perform machining with good machinability, the workpiece material should have the following properties/characteristics: (a) medium ductility; (b) reasonably low strain hardening exponent, shear strength, and fracture toughness; (c) presence of non-metallic inclusions that soften at high temperatures; (d) high thermal conductivity; and (e) a low metallurgical bond (adhesion) between cutting-tool and wokpiece.

Ductility directly affects the type of chip produced, which, in turn, affects surface finish. A brittle material (with a low ductility) may cause tool damage. On the other hand, very ductile materials tend to produce continuous chips that are difficult to control. Thus, medium ductility is desirable for good machinability. A work material with high thermal conductivity is helpful for dissipating heat to chips, resulting in cooler work during machining. It is important to avoid the embedding of very hard compounds (such as carbides, some oxides, and silicon) in the work material since these hard compounds accelerate tool wear.

Because steels are among the most commonly used engineering materials that are machined, the machinability of steel is generally improved by adding elements (e.g., lead, sulfur, etc.) or by heat treatment to change their properties. Besides free-machining steels, other good machinable materials include aluminum and its softer alloys, nodular cast irons, plastics, and nano-ceramics (Kalpakjian and Schmid, 2008).

1.6 INDUSTRIAL IMPORTANCE OF CALCULATIONS IN MACHINING

The calculations for various machining variables are of great technological importance in industrial practice. For example, the computation for spindle rotational speed in turning/drilling/milling enables a machinist to select the right spindle rpm in the machine tool. The computations for MRR in both traditional and NTM are also important for determining the efficiency and productivity. The calculation for the cutting time enables the machinist to complete the machining job well in time (Huda, 2018). The cutting time for a machining job also enables a machine shop manager to compute the labor cost in machining the job. The calculations for taper angle and tailstock offset in taper turning are important in the manufacture/machining of tapered shafts and other similar machine elements. The computation for MRR in NTM processes (e.g., AJM and ECM) is also important since the MRR enables a manufacturing manager to assess the efficiency and productivity of an NTM process.

The selection of an electric motor of correct horse-power (HP) to be attached to a machine tool is very important. In order to calculate the power required in machining, a machine-shop manager must calculate the cutting force by using a dynamometer as well as appropriate mathematical models. The cutting force calculations also enable a machinist to determine the degree of clamping required to hold a workpiece during machining. In gear manufacturing, the indexing calculations for both plain and differential indexing are of great industrial significance. For example, calculations for the number of revolutions of index crank and for differential change gear ratio enable a machinist/engineer to select the correct differential change gears in an indexing head attached to a milling machine. All the abovementioned aspects of mathematical modeling and calculations in machining are covered in this textbook.

QUESTIONS

1.1 Encircle the most appropriate answers for the following statements.
  a. Which type of machining process is the drilling operation?
    (i) abrasive machining, (ii) conventional machining, (iii) NTM, (iv) Micro-precision machining
  b. Which type of machining process is the grinding operation?
    (i) abrasive machining, (ii) conventional machining, (iii) NTM, (iv) Micro-precision machining
  c. Which type of machining process is the electron beam machining?
    (i) abrasive machining, (ii) conventional machining, (iii) NTM, (iv) Micro-precision machining
  d. Which type of machining process is the diamond turning operation?
    (i) abrasive machining, (ii) conventional machining, (iii) NTM, (iv) Micro-precision machining
  e. Which type of machining process is the honing operation?
    (i) abrasive machining, (ii) conventional machining, (iii) NTM, (iv) Micro-precision machining
  f. Which type of NTM process is abrasive jet machining (AJM)?
    (i) mechanical energy machining, (ii) thermal energy machining, (iii) electro-chemical machining, (iv) chemical machining.
  g. Which type of NTM process is electro-chemical grinding?
    (i) mechanical energy machining, (ii) thermal energy machining, (iii) electro-chemical machining, (iv) chemical machining.
  h. Which type of NTM process is electric discharge machining (EDM)?
    (i) mechanical energy machining, (ii) thermal energy machining, (iii) electro-chemical machining, (iv) chemical machining
1.2 Define the term machining. List the three fundamental cutting conditions and define them.
1.3 Illustrate the various types of machining processes with the aid of a classification chart.
1.4 Differentiate between roughing and finishing in machining.
1.5 a. What is meant by machinability?
  b. What properties are required for a material to be machinable.
  c. Give at least three examples of machinable materials.
1.6 Highlight the industrial importance of machining and its mathematical modeling with examples.

REFERENCES

Brinksmeier, E. (2014), Ultra precision machining. In: *CIRP Encyclopedia of Production Engineering*, pp. 1277–1280.

Huda, Z. (2018), *Manufacturing: Mathematical Models, Problems, and Solutions*, CRC Press, Boca Raton, FL.

Kalpakjian, S., Schmid, S. (2008), *Manufacturing Processes for Engineering Materials* (5th Edition), Pearson Education plc, London.

Kang, H.-J., Ahn, S.-H. (2007), Fabrication and characterization of micro-parts by mechanical micro-machining: Precision and cost estimation. *Proceedings of the Institution of Mechanical Engineers, Part B: Journal of Engineering Manufacture*, 22(2), 231–240.

Mativenga, P. (2014), Micro machining. In: *CIRP Encyclopedia of Production Engineering*, pp. 873–877.

Schneider, F., Das, J., Kirsch, B., Linke, B., Aurich, J.C. (2019), Sustainability in ultra precision and micro machining: A review, *International Journal of Precision Engineering and Manufacturing-Green Technology*, 6(3), 601–610.

2 Mechanics of Chip Formation

2.1 CHIP FORMATION IN MACHINING

We have learnt in the preceding chapter that conventional machining processes involve removal of material from a workpiece in the form of chips. The chip formation, in machining, is a localized shear deformation process that involves the removal of material from workpiece in the form of chip. The main cutting action in a conventional machining process involves shear deformation of the work material in a narrow region where the material is plastically deformed and slides off along the rake face of the cutting edge in the form of a chip (see Figure 2.1). As the chip is removed, a new surface is exposed. A chip consists of two sides: (a) the *shiny side* (flat and uniform side) that is in contact with the tool, due to frictional effects, and (b) the *jagged side*, which is the free workpiece surface, due to shear. The chip formation affects the surface finish, cutting forces, temperature, tool life, and dimensional tolerance (Liang and Shih, 2016; Huda, 2018).

2.2 ORTHOGONAL AND OBLIQUE MACHINING

The orientation of cutting edge of a wedge-shaped tool with reference to cutting velocity vector ($\vec{V}$) determines whether the machining is orthogonal or oblique (see Figure 2.2).

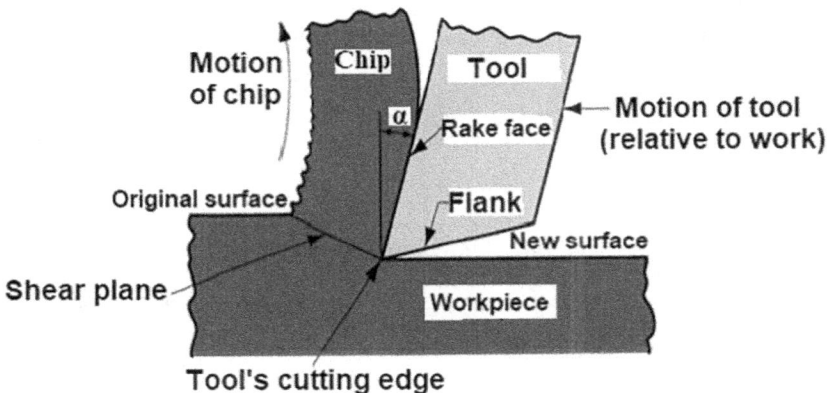

FIGURE 2.1 Chip formation in machining. (α = rake angle)

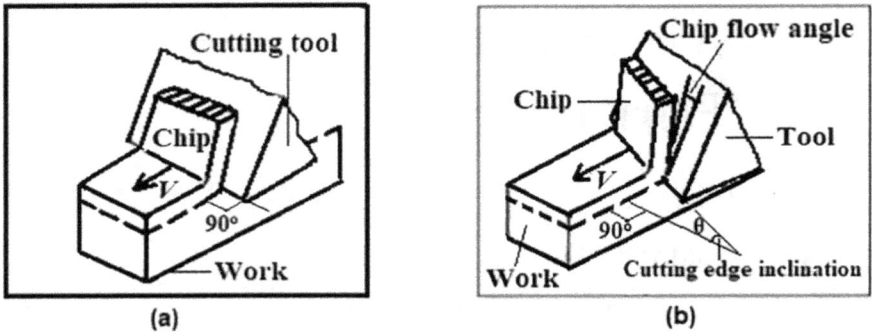

FIGURE 2.2 Orthogonal machining (a) and oblique machining (b)

2.2.1 ORTHOGONAL MACHINING

In *orthogonal machining,* the cutting edge is at right angle to the cutting velocity vector ($\vec{V}$), *i.e.*, the cutting edge is parallel to the uncut surface (Figure 2.2a). Thus, the cutting edge inclination is zero. Orthogonal cutting is represented by a 2D coordinate system because only two components of force (cutting force and thrust force) are considered. Examples of *orthogonal machining* include straight turning, lathe cutoff operation, surface broaching, peripheral milling, and the like.

2.2.2 OBLIQUE MACHINING

In *oblique machining*, the tool's cutting edge is inclined at an angle θ (other than 90°) to the cutting velocity vector (Figure 2.2b). *Oblique cutting* is represented by a 3D coordinate system because three components of force (cutting force, thrust force, and radial force) are considered. Examples of *oblique machining* include chamfering or oblique turning, drilling, grinding, and the like.

The difference in orthogonal machining and oblique machining is illustrated in Table 2.1.

TABLE 2.1
Difference in Orthogonal Machining and Oblique Machining

#	Orthogonal Machining	Oblique Machining
1	The tool's cutting edge is at right angle to the cutting velocity vector	The cutting edge is inclined at an angle other than 90° to the velocity vector
2	Represented by a 2D coordinate system	Represented by a 3D coordinate system
3	Examples include straight turning, lathe cutoff operation, surface broaching, peripheral milling, etc.	Examples include chamfering or oblique turning, drilling, grinding, etc.

2.3 ENGINEERING ANALYSIS OF CHIP FORMATION

In machine-tool practice, the knowledge and determination of chip formation parameters (e.g., chip thickness ratio, shear plane angle, etc.) are of great industrial importance as these parameters enable an engineer to compute forces in machining (see Chapter 3). Figure 2.3 illustrates the 2D model of machining that enables us to conduct engineering analysis of chip formation.

It is evident in Figure 2.3 that $t_o < t_c$, i.e., the uncut chip thickness is less than the deformed chip thickness. In the case of orthogonal machining, the uncut chip thickness equals the depth of cut, i.e., $t_o = d$, the uncut chip width equals the width of cut, w, and the uncut chip length equals the length of cut (Oxley, 1961). In micro-precision machining, the determination of the minimum uncut chip thickness (t_o) is important for optimizing the machining process (Rezaei et al., 2018). Figure 2.3 also illustrates that the cutting tool has two angles: (a) rake angle and (b) clearance angle. The rake angle (α) is the angle between the rake face of the tool and a line perpendicular to the workpiece. The clearance angle or the relief angle provides a small clearance between the newly machined work surface and the tool flank (see Figure 2.3).

Important chip formation parameters include chip thickness ratio (r), rake angle (α), shear plane angle (ϕ), shear strain (γ), and the like.

The chip thickness ratio (r) is given by:

$$r = \frac{t_o}{t_c} \qquad (2.1)$$

Since $t_o < t_c$, $r < 1$.

The shear plane angle (ϕ) can be computed by:

$$\phi = \tan^{-1}\left(\frac{r\cos\alpha}{1 - r\sin\alpha}\right) \qquad (2.2)$$

The shear strain (γ) is given by:

$$\gamma = \tan(\phi - \alpha) + \cot\phi \qquad (2.3)$$

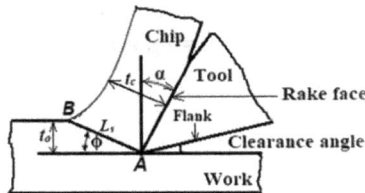

FIGURE 2.3 Chip formation parameters. (α = rake angle; L_s = length sheared; ϕ = shear plane angle)

The significance of Equations 2.1–2.3 is illustrated in Examples 2.1–2.4.

By reference to Figure 2.3, it can be mathematically shown that:

$$AB = \frac{t_o}{\sin\phi} = \frac{t_c}{\cos(\phi-\alpha)} \qquad (2.4)$$

By combining Equations 2.1 and 2.4, we obtain:

$$r = \frac{\sin\phi}{\cos(\phi-\alpha)} \qquad (2.5)$$

The significance of Equation 2.5 is illustrated in Example 2.4.

The shear plane area can be determined by reference to Figure 2.3 as follows. The shear plane area (A_s) is the product of the length sheared (L_s) and the width (w), *i.e.*:

$$A_s = L_s \cdot w \qquad (2.6)$$

By reference to Figure 2.3, we can write:

$$\sin\phi = \frac{t_o}{L_s} \qquad (2.7)$$

or

$$L_s = \frac{t_o}{\sin\phi} \qquad (2.8)$$

By combining Equations 2.6 and 2.8, we obtain:

$$A_s = \frac{t_o \cdot w}{\sin\phi} \qquad (2.9)$$

where A_s is the shear plane area, mm^2; t_o is the uncut chip thickness, mm; w is the width of cut, mm; and ϕ is the shear plane angle (see Example 2.5).

By using the equation of continuity, we may write:

$$t_o\, b_o\, L_o = t_c\, b_c\, L_c \qquad (2.10)$$

where b_o is the uncut chip width ($b_o = w$), L_o is the uncut chip length, b_c is the chip width, L_c is the chip length, and t_o and t_c have their usual meanings.

Since in most machining processes $b_o = b_c$, Equation 2.10 simplifies to:

$$t_o\, L_o = t_c\, L_c \qquad (2.11)$$

or

$$r = \frac{t_o}{t_c} = \frac{L_c}{L_o} \qquad (2.12)$$

By dividing both sides of Equation 2.11 by time t, we get:

$$t_o \left(\frac{L_o}{t} \right) = t_c \left(\frac{L_c}{t} \right)$$

or

$$t_o V = t_c V_c \tag{2.13}$$

$$\frac{V_c}{V} = \frac{t_o}{t_c} = r \tag{2.14}$$

where V is the cutting speed, and V_c is the chip speed. The significance of Equations 2.12–2.14 is illustrated in Examples 2.6 and 2.7.

2.4 THE EFFECTS OF RAKE ANGLE AND SHEAR PLANE ANGLE ON MACHINING

It is mentioned in Section 2.1 that machining involves the material removal/failure by shear deformation. We may thus deduce that the shear stress required in machining is equal to the shear strength of the work material. We learnt in Chapter 1 that the basic requirement for a good machinability is the minimum cutting and shearing forces. The rake angle and the shear plane angle play an important role in achieving this requirement. Figure 2.4 illustrates the effects of rake angle on the shear plane angle and area. It is evident in Figure 2.4 that an increase in rake angle α results in an increase of the shear plane angle ϕ, which in turn results in a lower shear plane area.

We know that the shear stress is given by:

$$\tau = \frac{F_s}{A_s} \tag{2.15}$$

where τ is the shear stress, MPa; F_s is the shear force, N (see Chapter 3); and A_s is the shear plane area, mm^2.

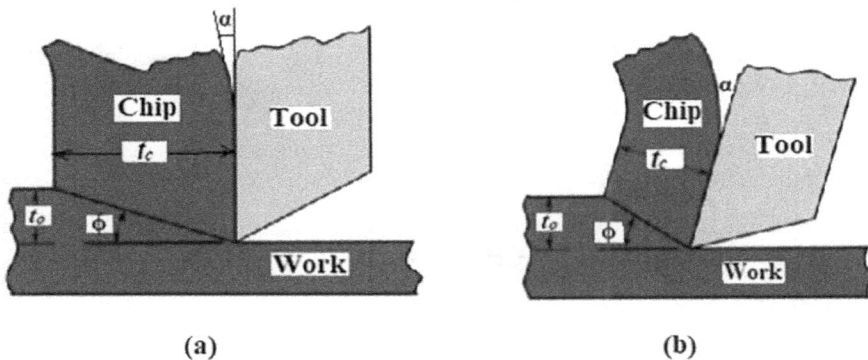

(a)   (b)

FIGURE 2.4   The effects of rake angle on shear plane angle and shear plane area

Rewriting Equation 2.15 for the shear force yields:

$$F_s = \tau A_s \qquad (2.16)$$

The significance of Equation 2.16 is illustrated in Example 2.8. Equation 2.16 also enables us to draw a useful conclusion of great industrial importance as follows.

A reference to Figure 2.4 indicates that an increase in the shear plane angle ϕ results in a decrease in shear plane area, A_s. By using Equation 2.16, we find that a lower A_s results in a lower shear force, which means easier machining. Thus, we may logically conclude by references to Figure 2.4 and Equation 2.16 that an increase in rake angle or shear plane angle results in easier machining.

2.5 MERCHANT'S EQUATION AND ITS INDUSTRIAL APPLICATION

We have established in the preceding section that an increase in the shear plane angle results in easier machining; it means that for easy machining, we must select a high shear plane angle that minimizes the shear force. The shear plane angle (ϕ) can be determined by Merchant's equation as follows:

$$\phi = 45° + \frac{\alpha}{2} - \frac{\beta}{2} \qquad (2.17)$$

where α and β are the rake angle and the friction angle, in degrees, respectively (see Example 2.9). A detailed explanation of friction angle is given in Chapter 3.

The Merchant's equation is very useful in achieving easy machining or higher shear plane angle. According to Merchant's equation (Equation 2.17), for easy machining, the rake angle should be higher, and the friction angle should be lower. The friction angle (β) is related to the coefficient of friction between the surfaces of the cutting tool and the work (μ) as follows:

$$\mu = \tan \beta \qquad (2.18)$$

The significance of Equations 2.17 and 2.18 is illustrated in Examples 2.10–2.12. Equations 2.17 and 2.18 are mathematical relationships of great industrial importance since they enable us to justify the use of lubricants for easy machining as follows. By using a lubricant during machining, the coefficient of friction μ is decreased, which results in a lower friction angle β (see Equation 2.16). According to the Merchant's equation (Equation 2.15), a lower friction angle results in a higher shear plane angle, which in turn results in easy machining (see Section 2.4).

2.6 TYPES OF CHIPS IN MACHINING

2.6.1 THE FOUR TYPES OF CHIPS

The formation of a particular chip in machining depends on the type of material being machined and the cutting conditions of the operation. There are four types of

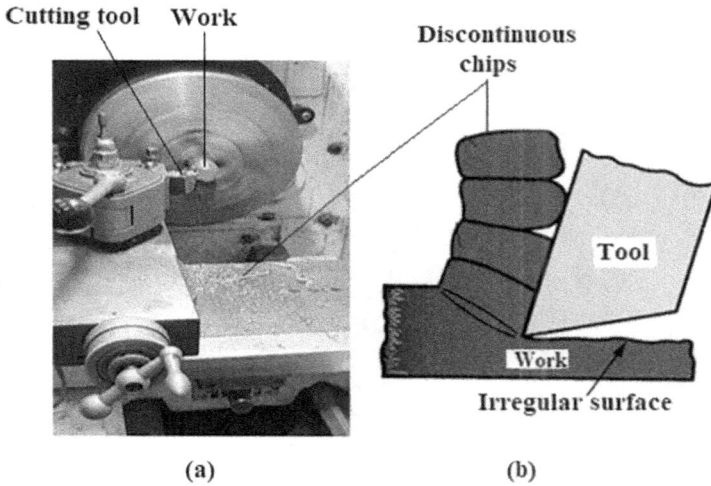

FIGURE 2.5 Photograph of discontinuous chips from a machined brass work (a); schematic illustration of discontinuous chips resulting in irregular machined surface (b)

chips that are commonly formed in machining: (a) discontinuous or segmented chip, (b) continuous chip, (c) continuous chip with built-up edge (BUE), and (d) serrated chip. Each of these types of chips is explained in the following sub-sections.

2.6.2 SEGMENTED OR DISCONTINUOUS CHIPS

The segmented or discontinuous chips refer to the segments of formed chips that may be firmly or loosely attached to each other during machining (see Figure 2.5a). In machine-shop practice, discontinuous chips are desirable because they do not cause any hindrance in machining. However, discontinuous chips result in irregular machined surface due to chip segmentation (see Figure 2.5b). The conditions of discontinuous chips formation include brittle work materials (e.g., cast irons), low cutting speeds, low rake angles, large feed, large depth of cut, and a high tool-chip friction.

2.6.3 CONTINUOUS CHIPS IN MACHINING

The continuous chips refer to long continuous chips, usually formed in turning ductile materials (see Figure 2.6a). A distinct advantage of continuous chip is that the machined surface has a good finish (see Figure 2.6b).

The operating conditions for obtaining continuous chip include high cutting speeds, high rake angles, small feeds, small depths of cut, a sharp cutting edge on the tool, and low tool-chip friction. There is a problem associated with a continuous chip since such a chip causes problems with regard to chip removing and/or tangling about the cutting tool. However, this problem can be overcome by using *chip breakers*. A chip breaker is clamped on the rake face of a cutting tool, as shown in

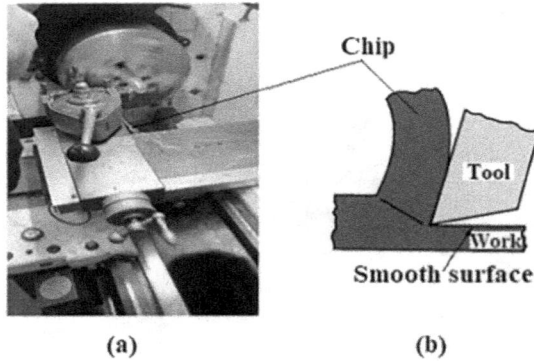

(a)   (b)

FIGURE 2.6   Photograph of continuous chip from a machined aluminum work (a); schematic illustration of discontinuous chips resulting in a good surface finish (b)

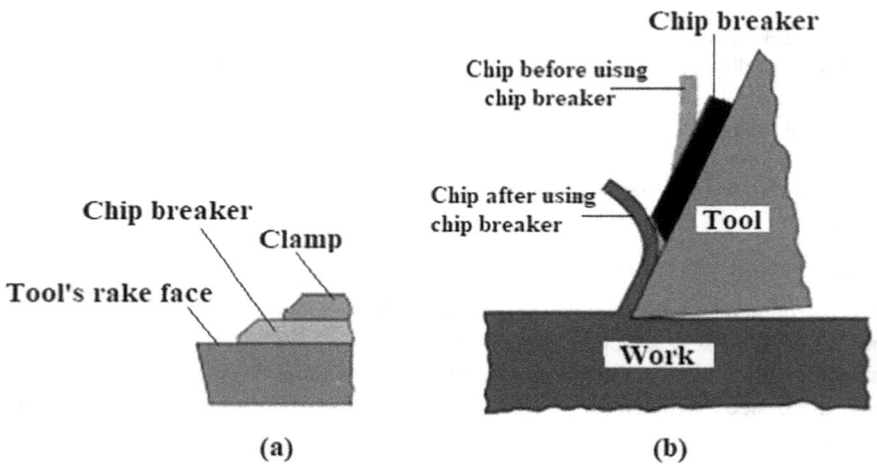

(a)   (b)

FIGURE 2.7   Chip breaker clamping (a); the action of chip breaker during machining (b)

Figure 2.7a. During machining, the chip breaker decreases the radius of curvature of the chip, resulting in fracture of chip, *i.e.*, the continuous chip is converted into segmented chips (see Figure 2.7b). Thus, the chip-tangling problem is solved while the work's surface finish remains good.

2.6.4 OTHER TYPES OF CHIPS

2.6.4.1 Continuous Chips With BUE

During machining with a ductile material at low-to-medium cutting speeds, tool-chip friction causes portions of continuous chip to adhere to the rake face of the cutting tool; this adherence results in the formation of BUE (see Figure 2.8a). Much of the

Serrated chips

High strain zone

Low strain zone

Chip

Tool

Built up edge

BUE particles

Tool

Work

(a)

(b)

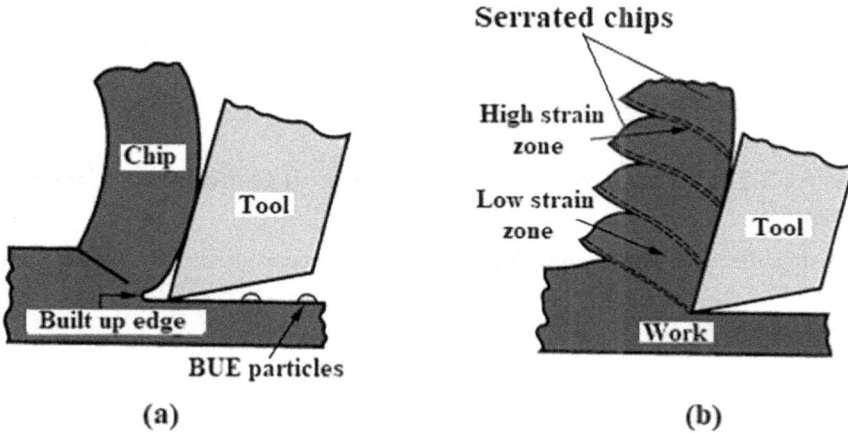

FIGURE 2.8 Continuous chips with built-up edge (a); and serrated chips (b)

detached BUE is carried away with the chip. Continuous chip with BUE results in a high surface roughness.

2.6.4.2 Serrated Chips

Serrated chips refer to a semi-continuous chip with saw-tooth appearance. Serrated chips are formed during machining difficult-to-machine metals (e.g., titanium alloys, nickel-base superalloys, and austenitic stainless steels) at higher cutting speeds.

2.7 CALCULATIONS—WORKED EXAMPLES ON MECHANICS OF CHIP FORMATION

EXAMPLE 2.1: CALCULATING THE CHIP THICKNESS RATIO AND SHEAR PLANE ANGLE

In an orthogonal machining operation, the cutting tool has a rake angle of 9°. The chip thickness before the cut is 0.60 mm, and the chip thickness after the cut is 1.10 mm. Calculate the (a) chip thickness ratio and (b) the shear plane angle.

Solution

$\alpha = 9°$, $t_o = 0.6$ mm, $t_c = 1.1$ mm, $r = ?$, $\phi = ?$

   a. By using Equation 2.1,

$$r = \frac{t_o}{t_c} = \frac{0.6\,\text{mm}}{1.1\,\text{mm}} = 0.545$$

   The chip thickness ratio = $r = 0.545$.

   b. By using Equation 2.2,

$$\phi = \tan^{-1}\left(\frac{r\cos\alpha}{1-r\sin\alpha}\right) = \tan^{-1}\left(\frac{0.545\cos 9°}{1-0.545\sin 9°}\right) = \tan^{-1}\left(\frac{0.545\times 0.988}{1-0.545\times 0.156}\right)$$

$$\phi = \tan^{-1}\left(\frac{0.538}{1-0.085}\right) = \tan^{-1}(0.588) = 30.4°$$

The shear plane angle $= \phi = 30.4°$.

EXAMPLE 2.2: CALCULATING THE SHEAR STRAIN FOR AN ORTHOGONAL MACHINING OPERATION

By using the data in Example 2.1, calculate the shear strain for the machining operation.

Solution

$\alpha = 9°$, $\phi = 30.4°$, $\gamma = ?$

By using Equation 2.3,

$$\gamma = \tan(\phi - \alpha) + \cot\phi = \tan(30.4 - 9) + \cot 30.4 = \tan 21.4° + \frac{1}{\tan 30.4°}$$

$$\gamma = 0.392 + \frac{1}{0.587} = 2.1$$

The shear strain $= 2.1$.

EXAMPLE 2.3: CALCULATING SHEAR STRAIN WITH THE AID OF SKETCH OF MACHINING OPERATION

By reference to Figure E-2.3, calculate the shear strain for the orthogonal machining operation.

Solution

From Figure E-2.3, $t_o = 0.4$ mm, $t_c = 0.7$ mm, $\alpha = 8°$, $\gamma = ?$

In order to determine the shear strain γ, we use Equations 2.1–2.3 as follows.

$$r = \frac{t_o}{t_c} = \frac{0.4}{0.7} = 0.57$$

$$\phi = \tan^{-1}\left(\frac{r\cos\alpha}{1 - r\sin\alpha}\right) = \tan^{-1}\left(\frac{0.57\cos 8°}{1 - 0.57\sin 8°}\right) = \tan^{-1}\left(\frac{0.57 \times 0.99}{1 - 0.57 \times 0.139}\right)$$

$$\phi = \tan^{-1}\left(\frac{0.56}{1 - 0.079}\right) = \tan^{-1}(0.608) = 31.3°$$

$$\gamma = \tan(\phi - \alpha) + \cot\phi = \tan(31.3 - 8) + \cot 31.3 = \tan 23.3° + \frac{1}{\tan 31.3°}$$

$$\gamma = 0.43 + \frac{1}{0.608} = 2.07$$

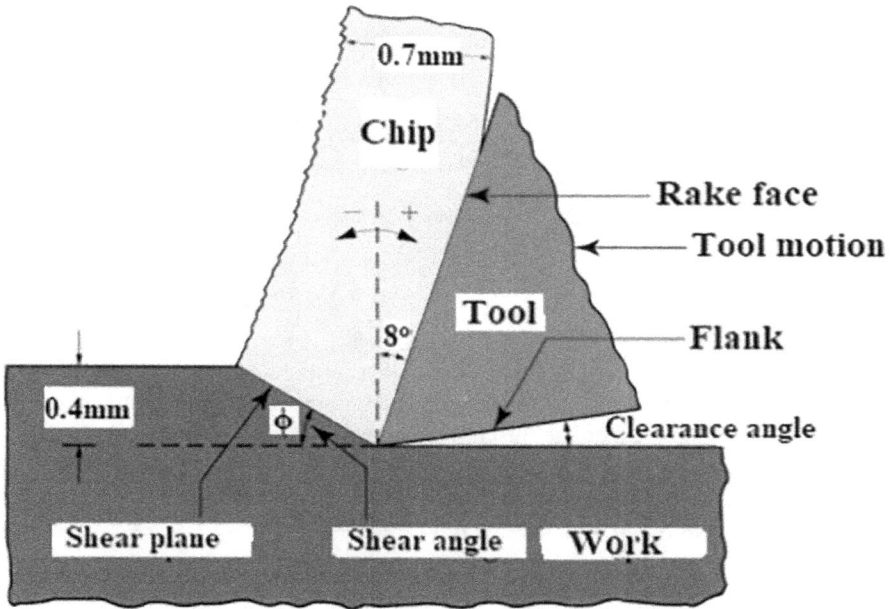

FIGURE E-2.3 Orthogonal machining with specified chip formation parameters

The shear strain = 2.07.

EXAMPLE 2.4: CALCULATING THE CHIP THICKNESS RATIO WHEN T_o AND T_c ARE UNKNOWN

The rake angle and shear plane angle for an orthogonal machining operation are 10° and 27°, respectively. Calculate the chip thickness ratio.

Solution
$\alpha = 10°$, $\phi = 27°$, $r = ?$
   By using Equation 2.5,

$$r = \frac{\sin \phi}{\cos(\phi - \alpha)} = \frac{\sin 27}{\cos(27 - 10)} = \frac{0.454}{0.956} = 0.474$$

The chip thickness ratio = 0.474.

EXAMPLE 2.5: CALCULATING THE SHEAR PLANE AREA IN MACHINING

In a machining operation, the width of the orthogonal cutting operation is 3.5 mm. The uncut chip thickness is 0.6 mm, and the shear plane angle is 28°. Calculate the shear plane area.

Solution
$w = 3.5$ mm, $t_o = 0.6$ mm, $\phi = 28°$, $A_s = ?$
   By using Equation 2.9,

$$A_s = \frac{t_o \cdot w}{\sin \phi} = \frac{0.6 \times 3.5}{\sin 28} = 4.47\,\text{mm}^2$$

The shear plane area = 4.47 mm².

EXAMPLE 2.6: COMPUTING CHIP THICKNESS AND Φ WHEN THE CHIP THICKNESS RATIO IS UNKNOWN

The end of a metal rod was orthogonally cut with a cutting tool of 18° rake angle. The length of the cut was 20 cm, and the length of the chip was 80 mm. The uncut chip thickness was 0.4 mm. Calculate the chip thickness and the shear plane angle.

Solution
$\alpha = 18°$, $L_o = 20$ cm $= 200$ mm, $L_c = 80$ mm, $d = t_o = 0.4$ mm, $t_c = ?$, $\phi = ?$
   By using Equation 2.12,

$$r = \frac{t_o}{t_c} = \frac{L_c}{L_o} = \frac{80}{200} = 0.4$$

$$t_c = \frac{t_o}{r} = \frac{0.4}{0.4} = 1\,\text{mm}$$

By using Equation 2.2,

$$\phi = \tan^{-1}\left(\frac{r\cos\alpha}{1-r\sin\alpha}\right) = \tan^{-1}\left(\frac{0.4 \times \cos 18}{1-0.4\sin 18}\right) = \tan^{-1}\left(\frac{0.4 \times 0.95}{1-0.4 \times 0.309}\right)$$

$$\phi = 23.4°$$

The chip thickness is 1 mm, and the shear plane angle is 23.4°.

EXAMPLE 2.7: CALCULATING THE CHIP SPEED IN ORTHOGONAL MACHINING

An orthogonal machining operation was carried out with a cutting speed of 120 m/min. The uncut chip thickness was 0.4 mm, and the thickness of the deformed chip was 0.5 mm. Calculate the chip speed.

Solution
$V = 120$ m/min, $t_o = 0.4$ mm, $t_c = 0.5$ mm, $V_c = ?$
   By using Equation 2.14,

$$\text{The chip speed} = V_c = V\left(\frac{t_o}{t_c}\right) = 120 = \left(\frac{0.4}{0.5}\right) = 96\,\text{m}/\text{min}$$

EXAMPLE 2.8: CALCULATING THE SHEAR FORCE WHEN THE WORK'S SHEAR STRENGTH IS GIVEN

The shear plane area in an orthogonal machining operation is 5 mm². Calculate the shear force required to machine the material if the shear strength of the work material is 250 MPa.

Solution

By reference to Section 2.4, the shear strength of the work material is equal to the shear stress required in machining. Thus,

$\tau = 250$ MPa $= 250$ N·mm^{-2}, $A_s = 5$ mm²; $F_s = ?$
By using Equation 2.14,

$$F_s = \tau A_s = 250 \times 5 = 1250 \, \text{N}$$

The shear force = 1250 N.

EXAMPLE 2.9: COMPUTING THE FRICTION ANGLE WHEN THE OTHER ANGLES ARE GIVEN

In an orthogonal machining operation, the rake angle is 7°, and shear plane angle is 25°. Calculate the friction angle.

Solution

$\alpha = 7°$, $\phi = 25°$, $\beta = ?$
By rewriting Merchant's equation (Equation 2.17), we obtain:

$$\beta = 2\left(45° + \frac{\alpha}{2} - \phi\right) = 90 + \alpha - 2\phi$$

$$\beta = 90 + 7 - (2 \times 25) = 97 - 50 = 47°$$

The friction angle = 47°.

EXAMPLE 2.10: CALCULATING THE COEFFICIENT OF FRICTION WHEN FRICTION FORCE IS UNKNOWN

By using the data in Example 2.9, calculate the coefficient of friction between the surfaces of the cutting tool and the work.

Solution

By using Equation 2.18,
The coefficient of friction = $\mu = \tan\beta = \tan 47° = 1.07$.

EXAMPLE 2.11: COMPUTING THE DEFORMED CHIP THICKNESS, SHEAR PLANE ANGLE, FRICTION ANGLE, COEFFICIENT OF FRICTION, AND SHEAR STRAIN

In an orthogonal cutting operation, the rake angle is −4°, the uncut chip thickness is 0.18 mm, and the width of the cut is 3.8 mm. The chip thickness ratio is 0.37. Determine (a) the deformed chip thickness, (b) the shear plane angle, (c) the friction angle, (d) the coefficient of friction, and (e) the shear strain.

Solution

$\alpha = -4°$, $r = 0.37$, $t_o = 0.18$ mm, $w = 3.8$ mm, $t_c = ?$, $\phi = ?$, $\beta = ?$, $\mu = ?$, $\gamma = ?$

a. By rewriting Equation 2.1 for t_c, we obtain:

$$t_c = \frac{t_o}{r} = \frac{0.18\,\text{mm}}{0.37} = 0.486\,\text{mm}$$

The deformed chip thickness $= t_c = 0.486$ mm.

b. By using Equation 2.2,

$$\phi = \tan^{-1}\left(\frac{0.37\cos(-4)}{1-(0.37\times\sin-4)}\right) = \tan^{-1}\left(\frac{0.37\times0.997}{1-(0.37\times-0.07)}\right) = \tan^{-1}\left(\frac{0.369}{1+0.026}\right)$$

$$\phi = \tan^{-1}\left(\frac{0.369}{1.026}\right) = \tan^{-1}(0.36) = 19.8°$$

The shear plane angle $= \phi = 19.8°$.

c. $\alpha = -4°$, $\phi = 19.8°$

By rewriting Merchant's equation (Equation 2.17), we obtain:

$$\beta = 2\left(45° + \frac{\alpha}{2} - \phi\right) = 90 + \alpha - 2\phi$$

$$\beta = 90 + (-4) - (2\times19.8) = 90 - 4 - 39.6 = 46.4°$$

The friction angle $= \beta = 46.4°$.

d. By using Equation 2.18,

$$\mu = \tan\beta = \tan 46.4° = 1.05$$

The coefficient of friction $= \mu = 1.05$.

e. $\alpha = -4°$, $\phi = 19.8°$, $\gamma = ?$

By using Equation 2.3,

$$\gamma = \tan(19.8 + 4) + \cot 19.8 = \tan 23.8 + \cot 19.8 = \tan 23.8° + \frac{1}{\tan 19.8°}$$

$$\text{Shear strain} = \gamma = 0.441 + \frac{1}{0.36} = 3.22$$

EXAMPLE 2.12: COMPUTING THE SHEAR PLANE ANGLE WHEN THE FRICTION COEFFICIENT IS DOUBLED AND COMMENTING ON THE MACHINABILITY

In orthogonal cutting, the rake angle is 26°, and the coefficient of friction is 0.22. If the coefficient of friction is doubled, compute the shear plane angle in each case. Comment on the effect of change in friction coefficient on machinability.

Solution

Case (a)

$\alpha = 26°$, $\mu = 0.22$, $\phi = ?$

$$\tan \beta = \mu$$

$$\beta = \tan^{-1} \mu = \tan^{-1} 0.22 = 12.4°$$

$$\phi = 45° + \frac{\alpha}{2} - \frac{\beta}{2} = 45° + \frac{26}{2} - \frac{12.4}{2} = 45 + 13 - 6.2 = 51.8°$$

Case (b)

$\alpha = 26°$, $\mu = 0.22 \times 2 = 0.44$, $\phi = ?$

$$\beta = \tan^{-1} \mu = \tan^{-1} 0.44 = 23.7°$$

$$\phi = 45° + \frac{\alpha}{2} - \frac{\beta}{2} = 45° + \frac{26}{2} - \frac{23.7}{2} = 45 + 13 - 11.87 = 46.13°$$

It is obvious that by doubling the friction coefficient, the shear plane angle decreases; this means that machinability becomes poorer, *i.e.*, difficult machining.

QUESTIONS AND PROBLEMS

2.1 Encircle the most appropriate answers for the following statements:
   a. Which type of machining process is drilling?
      (i) non-traditional machining, (ii) orthogonal machining, (iii) oblique machining.
   b. Which type of machining process is straight turning?
      (i) non-traditional machining, (ii) orthogonal machining, (iii) oblique machining.
   c. Which type of chip is formed in machining of copper at high cutting speeds?
      (i) segmented chip, (ii) continuous chip, (iii) continuous chip with built-up edge, and (iv) serrated chip.
   d. Which type of chip is formed in machining of superalloys at high cutting speeds?
      (i) segmented chip, (ii) continuous chip, (iii) continuous chip with built-up edge, and (iv) serrated chip
   e. Which type of chip is formed in machining of cast iron at low cutting speeds?

(i) segmented chip, (ii) continuous chip, (iii) continuous chip with built-up edge, and (iv) serrated chip

f. Which type of chip is formed in machining a ductile material at low-to-medium speeds?
(i) segmented chip, (ii) continuous chip, (iii) continuous chip with built-up edge, and (iv) serrated chip

2.2 Explain the main cutting action in conventional machining with the aid of a diagram.

2.3 Distinguish between orthogonal and oblique machining with the aid of diagrams giving examples for each type of machining.

2.4 Diagrammatically illustrate the effect of increase in rake angle on shear plane area.

2.5 Draw diagrams showing the four types of chips in machining.

2.6 By using Merchant's equation, explain that the use of lubricants results in easier machining.

P2.7 In an orthogonal machining operation, the chip thickness before and after the cut are 0.39 mm and 0.68 mm, respectively. The cutting tool has a rake angle of 11°. Calculate: (a) the chip thickness ratio and (b) the shear plane angle.

P2.8 By using the data in P2.7, calculate the shear plane area if the width of the cutting operation is 3 mm.

P2.9 By using the data in P2.7, calculate the shear strain for the machining operation.

P2.10 By using the data in P2.7, calculate the friction angle in the machining operation.

P2.11 By using the data in P2.7, calculate the coefficient of friction.

P2.12 A 25-cm long metal workpiece was orthogonally machined with a cutting tool of 16° rake angle. The length of chip was 100 mm. The depth of cut was 0.6 mm. Calculate the chip thickness and the shear plane angle.

REFERENCES

Huda, Z. (2018), *Manufacturing: Mathematical Models, Problems, and Solutions*, CRC Press, Boca Raton, FL.

Liang S., Shih, A.J. (2016), *Analysis of Machining and Machine Tools*, Springer, Berlin.

Rezaei, H., Sadeghi, M.H., Budak, E. (2018), Determination of uncut chip thickness under various machining conditions during micro-milling of Ti-6Al-4V, *International Journal of Advanced Manufacturing Technology*, 95, 1617–1634.

Oxley, P.L.B. (1961), Mechanics of metal cutting, *International Journal of Machine Tool Design and Research*, 1(1–2), 89–97.

3 Forces and Power in Machining

3.1 IMPORTANCE OF CUTTING FORCES AND POWER

It has been introduced in the preceding chapter that the main cutting action in a conventional machining process involves shear deformation of the work material in a narrow region leading to failure/removal of material. In order to produce the shear deformation of the work material, a reasonably high shear force is required. In fact, a number of forces act on the work/tool during machining; these forces include the cutting force, the thrust force, the friction force, the normal force, the shear force, and the like. The determination and control of these forces and power are very important in machine shop practice (Cica et al., 2015). For example, the determination of the cutting force enables us to compute the cutting power; the latter enables us to assess electricity consumption costs and to select machine tool electric motor of right power (in kW or HP). The cutting force determination is also helpful in assessing the forces acting on the work holding clamp so as to select a stiff material for the device thereof. The determination of the thrust force is also important since the tool holder device, work holding clamp, and machine tool must be stiff enough to support the thrust force with minimal deflections. The control of thrust force is also important because if the thrust force is too high, the cutting tool will be pushed away from workpiece, which will reduce the depth of cut and dimensional accuracy.

3.2 FORCES ON WORK, TOOL, AND CHIPS DURING MACHINING

3.2.1 FORCES ON WORKPIECE

The cutting force $(\vec{F_c})$ acts on the workpiece and supplies the energy required for cutting. The cutting force acts in the direction of the cutting velocity $(\vec{V})$ as shown in Figure 3.1(a). In straight turning, the cutting force acts tangential to the workpiece [see Figure 3.1(b)]. The cutting force $\vec{F_c}$ needed to cut a chip area of 1 mm² and a thickness of 1 mm is called the *specific cutting force*. Another force that acts on the work is the thrust force $\vec{F_t}$, which acts in a direction normal to $\vec{F_c}$ [see Figure 3.1(a)]. In straight turning, the thrust force is called the feed force $(\vec{F_f})$ [see Figure 3.1(b)], the depth of cut = uncut chip thickness = t_o, and width of cut = feed = f.

3.2.2 FORCES ON TOOL/CHIP

In the preceding paragraph, the forces acting on the workpiece have been discussed. Besides the workpiece, the tool and the chip are also acted upon by some forces during machining (Astakhov, 1998). For example, the friction force $(\vec{F})$ acts along the

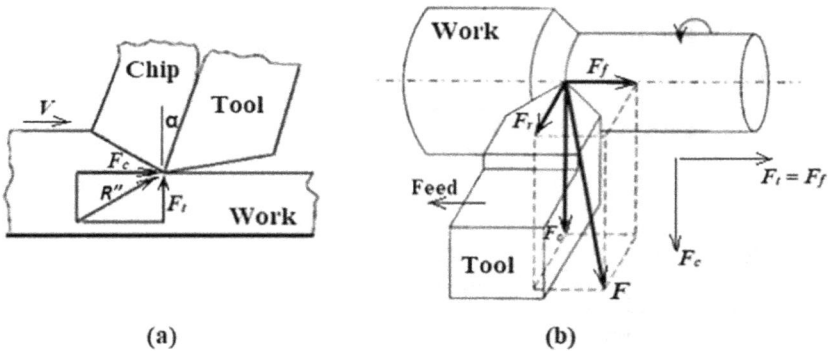

(a)   (b)

FIGURE 3.1 The cutting force and thrust force in orthogonal machining (a), the forces acting on the work in straight turning (b). (F_c = cutting force, F_t = thrust force, F_f = feed force, F_r = radial force)

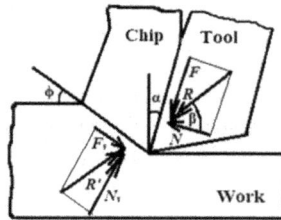

FIGURE 3.2 Forces acting on tool/chip during machining. (α = rake angle, β = friction angle, ϕ = shear plane angle, F = friction force, N = normal force, F_s = shear force, N_s = normal to shear force)

tool–chip interface; the force normal to $\vec{F}$ is called the normal force ($\vec{N}$), as shown in Figure 3.2. Thus the vector addition of $\vec{F}$ and $\vec{N}$ yields the resultant force ($\vec{R}$) that acts on the tool face. Additionally, the shear force ($\vec{F_s}$) acts along the shear plane; the force acting normal to $\vec{F_s}$, is called the normal to shear force ($\vec{N_s}$) (see Figure 3.2). It is evident in Figure 3.2 that in order to keep a balance during machining, the force $\vec{R'}$ must be equal in magnitude and opposite in direction to the resultant force $\vec{R}$; additionally the forces $\vec{R'}$ and $\vec{R}$ must be collinear.

3.3 MEASURING THE CUTTING FORCE AND THE THRUST FORCE

The cutting force F_c and the thrust force F_t can be measured by use of a device: *dynamometer* (see Figure 3.3). The working principle of a *dynamometer* is based on four elastic octagonal rings on which strain gauges are mounted and necessary connections are made to form the Wheatstone bridges (Yaldız and Ünsaçar, 2006). The strain (ε), induced by the cutting force, changes the electrical resistance R of the strain gauges. The change in the resistance of the gauges connected in a Wheatstone bridge produces voltage output ΔV through a strain measuring bridge (SMB).

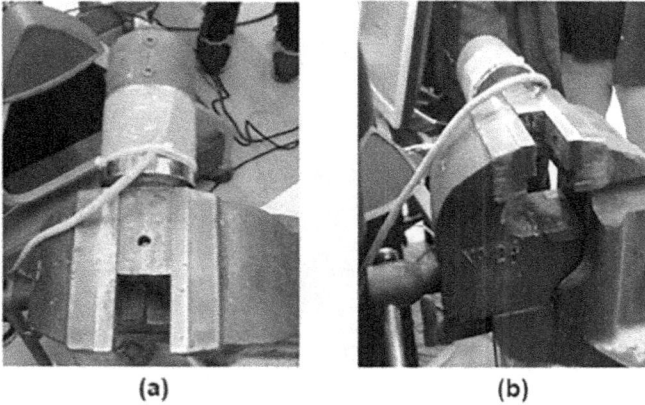

(a)   (b)

FIGURE 3.3   Dynamometer oriented/clamped to measure the cutting force (a) and the thrust force (b)

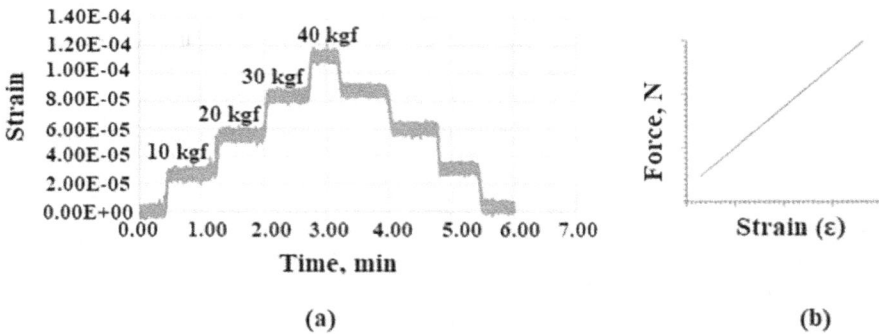

(a)   (b)

FIGURE 3.4   Calculating the constant a and b by using a strain–time plot (a) and calibration curve (b)

The load or force applied to the dynamometer is related to the strain, ε, by:

$$F = a + b \cdot \varepsilon \qquad (3.1)$$

where F is the force, N; a and b are constants. The constants a and b can be determined by the *calibration curve* (see Figure 3.4). The calibration curve may be plotted based on the strains obtained by applying loads of (say) 10, 20, 30, and 40 kgf and by noting the resulting strains (see Figure 3.4a). The use of the computer-based Least Squares Fitting software enables us to calculate the constants a and b in Equation 3.1 by use of the data acquisition (DAQ) system (see the calibration curve in Figure 3.4b) (see Example 3.1).

Once the constants a and b have been computed, the cutting force F_c (or thrust force F_t) can be calculated by determining the actual strain (ε) during the machining operation. For this purpose, the dynamometer cutting tool system is clamped in the

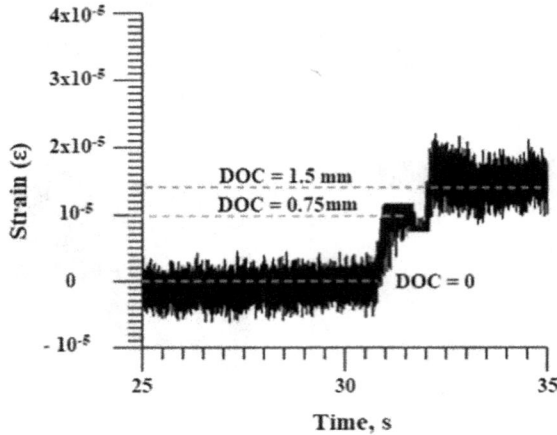

FIGURE 3.5 A typical output of DAQ during turning operation with different DOC for determining F_c

machine tool (*e.g.*, a lathe), and the resulting (actual) strains during machining are recorded. Accordingly, various values of strains at the different depth of cut (DOC) are obtained and recorded, as shown in Figure 3.5. These strain (ε) values are substituted in Equation 3.1 to obtain F_c (or F_t, as the case may be) (see Example 3.2).

3.4 MATHEMATICAL MODELS FOR FORCES IN MACHINING

The cutting force F_c and the thrust force F_t, in orthogonal machining, can be measured by a dynamometer; the other unknown forces (friction force F, normal force N, shear force F_s, and normal to shear force, N_s) can be determined by developing the mathematical models (Huda, 2018). By reference to Figures 3.1 and 3.2, the decomposition of $\vec{R}$ at the tool–chip interface results in:

$$F = F_c \sin \alpha + F_t \cos \alpha \qquad (3.2)$$

$$N = F_c \cos \alpha - F_t \sin \alpha \qquad (3.3)$$

The significance of Equations 3.2 and 3.3 is illustrated in Example 3.3.

By reference to Figures 3.1 and 3.2, the decomposition of $\vec{R'}$ at the shear plane results in:

$$F_s = F_c \cos \phi - F_t \sin \phi \qquad (3.4)$$

$$N_s = F_c \sin \phi + F_t \cos \phi \qquad (3.5)$$

The significance of Equations 3.4 and 3.5 is illustrated in Example 3.4. Once the shear force F_s has been calculated by using Equation 3.4, the shear stress can be computed by using Equation 2.13 (see Example 3.5).

The friction force F is related to the normal force N by:

$$F = \mu N \tag{3.6}$$

where μ is the coefficient of friction between tool and chip interface (see Example 3.6).

3.5 MERCHANT'S FORCE CIRCLE

It is learnt in Section 3.3 that the cutting force F_c and the thrust force F_t can be measured by use of a dynamometer. Once the values of F_c and F_t are known, the other cutting forces (F, N, F_s, and N_s) can be determined by using Equations 3.2–3.5. Another way of determining the friction force (F), the normal force (N), the shear force (F_s), and N_s is the geometrical method – Merchant's Force Circle method (Kesavan and Ramnath, 2010). The first step in drawing the Merchant's force circle is to choose a suitable scale and then representing the known forces F_c and F_t by lines indicating directions, as shown in Figure 3.1(a); the procedure of drawing the Merchant's force circle and the measurements of the forces F, N, F_s, and N_s and the friction angle β are illustrated in Examples 3.7 and 3.8.

3.6 POWER AND ENERGY IN MACHINING

Cutting Power. The *cutting power* is the power input required for cutting; it is given by:

$$P_c = F_c V \tag{3.7}$$

where P_c is the cutting power, W; F_c is the cutting force, N; and V is the cutting velocity, m/s (see Example 3.10).

Gross Motor Power. The *gross motor power* (in W) is the gross power of machine tool motor; it can be calculated by:

$$\text{Gross motor power} = \frac{F_c \cdot V}{\eta} \tag{3.8}$$

where η is the machine tool motor efficiency (see Example 3.11).

Unit Power or Specific Energy. The *unit power* or the *specific energy* (P_u in W·s/mm³) is the cutting power per unit volume rate of the metal cut; it is given by (Groover, 2016):

$$P_u = \frac{P_c}{MRR} = \frac{F_c \cdot V}{MRR} \tag{3.9}$$

where V is the cutting speed, m/s; and MRR is the metal removal rate, mm³/s.

The unit power requirements for machining of various alloys are presented in Table 3.1.

TABLE 3.1
The Unit Powers for Conventional Machining of Various Alloys

Alloys	Carbon Steel	Stainless Steel	Cast Iron	Copper Alloys	Aluminum Alloys	Magnesium Alloy	Titanium Alloy
P_u, W·s/mm³	2.7–9.3	3.0–5.2	1.6–5.5	1.4–3.3	0.4–1.1	0.4–0.6	3.0–4.1

The significance of Equation 3.9 and Table 3.1 is illustrated in Example 3.12.

The material removal rate (MRR) is related to the cutting speed (V) and the width of cut (w) by:

$$MRR = w\,t_o\,V \tag{3.10}$$

where MRR is expressed in mm³/s; w is in mm; t_o is the uncut chip thickness, mm; and V is in mm/s. The combination of Equations 3.9 and 3.10 yields:

$$P_u = \frac{F_c \cdot V}{MRR} = \frac{F_c \cdot V}{w \cdot t_o \cdot V} = \frac{F_c}{w \cdot t_o} \tag{3.11}$$

Specific Energy for Friction. Friction force plays an important role in machining (see Figure 3.2). The *specific energy for friction* can be expressed by rewriting Equation 3.9 as follows:

$$U_f = \frac{F \cdot V_c}{MRR} = \frac{F \cdot V_c}{w\,t_o\,V} \tag{3.12}$$

where U_f is the specific energy for friction, W·s/mm³; F is the friction force, N; and V_c is the chip speed, m/s.

By combining Equations 2.14 and 3.12, we obtain:

$$U_f = \frac{F \cdot r}{w\,t_o} \tag{3.13}$$

where r is the chip thickness ratio.

Specific Energy for Shearing. Machining involves shear deformation of the work material in a narrow region leading to failure/removal of material. The *specific energy for shearing* (U_s) is:

$$U_s = \frac{F_s \cdot V_s}{MRR} = \frac{F_s \cdot V_s}{w\,t_o\,V} \tag{3.14}$$

where U_s is the specific energy for shearing, W·s/mm³; F_s is the shear force, N; V_s is the shearing speed, m/s; w is in width, mm; t_o is in mm; and V is in mm/s.

Total Specific Energy. The total specific energy U is the sum of the specific energy for friction U_f and the specific energy for shear U_s i.e.,

$$U = U_f + U_s = P_u \qquad (3.15)$$

The significance of Equations 3.13–3.15 is illustrated in Examples 3.13–3.16.

3.7 TEMPERATURE IN MACHINING

Cutting temperature is an important factor in machining operations. The energy dissipated in machining is converted to heat, which in turn raises the temperature. A significant rise in cutting temperature is detrimental since it may result in tool wear, tool failure, and an inferior surface quality. The temperature, during machining, mainly depends on the cutting conditions and work material properties. The mean temperature in orthogonal machining operations is given by:

$$T = \frac{1.2 Y_f}{c} \left(\frac{V t_o}{K} \right)^{0.33} \qquad (3.16)$$

where T is the mean temperature at the tool–chip interface, K; Y_f is the flow strength of the work material, kPa; V is the cutting speed; t_o is the depth of cut, cm; c is the volumetric specific heat of the work material, cm·kg/cm^3·°C; and K is the thermal diffusivity of the work material, cm^2/s (Kalpakjian and Schmid, 2008) (see Example 3.17). The temperature at the tool–chip interface can be measured by use of a thermocouple that is embedded either in the tool or the workpiece.

The mean temperature at the tool–chip interface during turning operation using high-speed steel (HSS) tool is related to the cutting speed and the feed by:

$$T = C V^{0.5} f^{0.375} \qquad (3.17)$$

where T is the mean temperature, K; C is a constant; V is the cutting speed; and f is the feed.

3.8 CALCULATIONS – WORKED EXAMPLES ON CUTTING FORCES AND POWER

EXAMPLE 3.1: DRAWING A CALIBRATION CURVE AND CALCULATING THE CONSTANTS A AND B

Refer to the graphical plot [Figure 3.4(a)] obtained by loading/unloading by using a dynamometer oriented for measuring the cutting force F_c. By using the data in the plot, obtain the calibration curve, and hence calculate the constants a and b for use in Equation 3.1.

Solution

By using $F = W = mg$, we obtain Table E-3.1 for the various loads and the corresponding strains.

By using Table E-3.1, we obtain the force–strain (F – ε) plot: calibration curve (Figure E-3.1).

From the calibration curve (Figure E-3.1), the slope is determined to be $b = 3.27 \times 10^6$. Accordingly, by substituting the value of b and the data in Table E-3.1 (or Figure E-3.1) in Equation 3.1, we obtain:

$$98.1 = a + \left[\left(3.27 \times 10^6\right) \times \left(2.5 \times 10^{-5}\right) \right]$$

or

$$a = 16.35.$$

Hence, the calibration curve yields the constants: $a = 16.35$ and $b = 3.27 \times 10^6$.

TABLE E-3.1

Force Strain Data Obtained from the Strain–Time Plot

Force, N	98.1	196.2	294.3	392.4
Strain (ε)	2.5×10^{-5}	5.5×10^{-5}	8.5×10^{-5}	11×10^{-5}

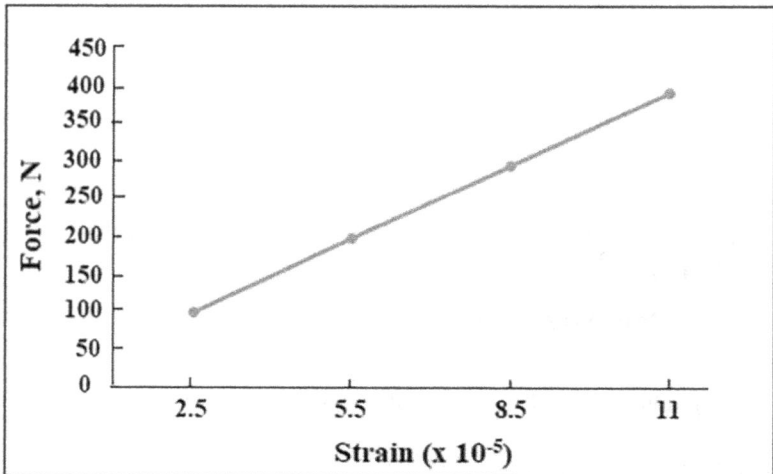

FIGURE E-3.1   Force–strain plot – *calibration curve*

EXAMPLE 3.2: CALCULATING THE CUTTING FORCE FROM THE OUTPUTS FROM DYNAMOMETER

Calculate the cutting force for various strains at the corresponding depths of cut (DOC) (as shown in Figure 3.5) in machining of aluminum. Use the constants computed in Example 3.1.

Solution

From Figure 3.5: (a) for DOC = 0, ε = 0, (b) for DOC = 0.75, ε = 10^{-5}, (c) for DOC = 1.5, ε = 1.5 × 10^{-5}.

a. For DOC = 0, ε = 0, a = 16.35 and b = 3.27 × 10^6

By using Equation 3.1,

$$\text{Cutting force} = F_c = a + b \cdot \varepsilon = 16.35 + \left(3.27 \times 10^6 \times 0\right) = 16.35\,\text{N}$$

b. For DOC = 0.75 mm, ε = 10^{-5}, a = 16.35 and b = 3.27 × 10^6

$$F_c = a + b \cdot \varepsilon = 16.35 + \left(3.27 \times 10^6 \times 10^{-5}\right) = 16.35 + 32.7 = 49\,\text{N}$$

c. For DOC = 1.5 mm, ε = 1.5 × 10^{-5}, a = 16.35 and b = 3.27 × 10^6

$$F_c = a + b \cdot \varepsilon = 16.35 + \left(3.27 \times 10^6 \times 1.5 \times 10^{-5}\right) = 16.35 + 49.05 = 65\,\text{N}$$

EXAMPLE 3.3: CALCULATING THE FRICTION FORCE AND THE NORMAL FORCE

The cutting force and the thrust force in an orthogonal cutting operation are 1400 and 950 N, respectively. The rake angle is 7°. Calculate the (a) friction force and (b) normal force.

Solution

α = 7°, F_c = 1400 N, F_t = 950 N, F = ?, N = ?

a. By using Equation 3.2,

$$\text{Friction force} = F = F_c \sin\alpha + F_t \cos\alpha = \left(1400\sin 7°\right) + \left(950\cos 7°\right)$$
$$= 170.6 + 943 = 1113.6\,\text{N}$$

b. By using Equation 3.3,

$$\text{Normal force} = N = F_c \cos\alpha - F_t \sin\alpha = \left(1400\cos 7°\right) - \left(950\sin 7°\right)$$
$$= 1389.5 - 115.7 = 1273.8\,\text{N}$$

EXAMPLE 3.4: CALCULATING THE SHEAR FORCE AND THE NORMAL TO SHEAR FORCE

In an orthogonal cutting operation, the shear plane angle is 24°. The cutting force is 1300 N, and the thrust force is 950 N. Calculate: (a) shear force and (b) normal to the shear force.

Solution
$F_c = 1300$ N, $F_t = 950$ N, $\phi = 24°$, $F_s = ?$, $N_s = ?$
  a. By using Equations 3.4 and 3.5,

$$\text{Shear force} = F_s = F_c \cos\phi - F_t \sin\phi = (1300\cos 24°) - (950\sin 24°)$$
$$= 801.2\,\text{N}$$

$$\text{Normal to shear force} = N_s = F_c \sin\phi + F_t \cos\phi$$
$$= (1300\sin 24°) + (950\cos 24°) = 1396.6\,\text{N}$$

EXAMPLE 3.5: CALCULATING THE SHEAR STRENGTH WITHOUT USING SHEAR TESTING MACHINE

In a straight turning operation, the width of the orthogonal cutting operation is 3.5 mm. The depth of cut is 0.6 mm, and the shear plane angle is 28°. The shear force is 1500 N. Calculate the shear strength.

Solution
$w = 3.5$ mm, $t_o =$ depth of cut $= 0.6$ mm (see Section 3.2.1), $\phi = 28°$, $F_s = 1500$ N, $\tau = ?$
  By using Equation 2.9,

$$A_s = \frac{t_o \cdot w}{\sin\phi} = \frac{0.6 \times 3.5}{\sin 28} = 4.47\,\text{mm}^2$$

By using Equation 2.13,

$$\text{Shear stress} = \tau = \frac{F_s}{A_s} = \frac{1500}{4.47} = 335.57\,\text{N}$$

In machining, the shear strength = the shear stress = 335.57 N.

EXAMPLE 3.6: CALCULATING THE COEFFICIENT OF FRICTION BY USING FORCES IN MACHINING

By using the data in Example 3.3, compute the coefficient of friction between the tool and chip interface.

Solution
Friction force $= F = 1113.5$ N, Normal force $= N = 1273.8$ N.
  By using modified form of Equation 3.6,

$$\text{The coefficient of friction} = \mu = \frac{F}{N} = \frac{1113.5}{1273.8} = 0.874.$$

EXAMPLE 3.7: DETERMINING F, N, F_s, AND N_s BY DRAWING THE MERCHANT'S FORCE CIRCLE

In an orthogonal machining operation, the cutting force and the thrust force are 1560 and 1270 N, respectively. The rake angle is 10° and the shear plane angle is 25°. Determine the friction force, the normal force, the shear force, and the normal to shear force by drawing the Merchant's Force Circle.

Solution

$F_c = 1560$ N, $F_t = 1270$ N, $\alpha = 10°$, $\phi = 25°$, $F = ?$, $N = ?$, $F_s = ?$, $N_s = ?$

We may conveniently choose a scale 1 cm $\equiv$ 500 N. The steps to be followed are given below:

I. Draw a line $\overrightarrow{OA}$ of length 3.12 cm representing $F_c = 1560$ N (see Figure E-3.7);
II. Draw a perpendicular line $\overrightarrow{AB}$ of length 2.54 cm representing $F_t = 1270$ N;
III. Join $\overrightarrow{OB} = \vec{R}$
IV. By using a compass, draw a circle with the mid-point of OB as the center.
V. By using a protractor, draw a line OC using the shear plane angle, $\phi = 25°$; the line $\overrightarrow{OC}$ represents the shear force, F_s. Measure the length OC giving 1.8 cm = 900 N = F_s.
VI. Join $\overrightarrow{CB} = N_s$. Measure the length CB giving 3.6 cm = 1800 N = N_s.
VII. Draw a vertical line passing through the point "O" and show the rake angle $\alpha = 10°$ as shown in Figure E-3.7). Draw the line $\overrightarrow{OD}$ representing the friction force, $\vec{F}$. Measure the length OD giving 3.16 cm = 1550 N = F.
VIII. Join $\overrightarrow{DB} = \vec{N}$. Measure the length DB giving 2.6 cm = 1300 N = Normal force.

Shear force, $F_s = 900$ N, $N_s = 1800$ N, Friction force, $F = 1550$ N, Normal force, $N = 1300$ N.

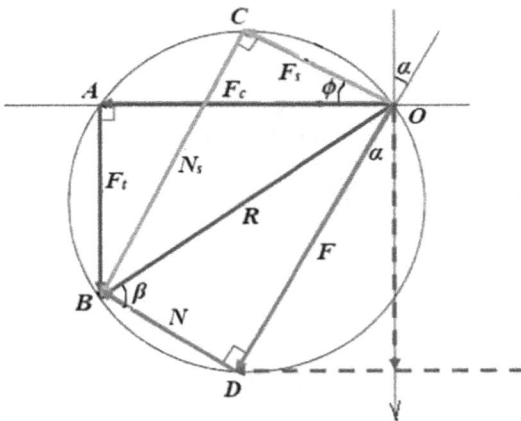

FIGURE E-3.7   Merchant's force circle. (not according to the chosen scale)

EXAMPLE 3.8: DETERMINING THE FRICTION ANGLE BY USING MERCHANT'S FORCE CIRCLE

By reference to the Merchant's force circle drawn in Example 3.7, measure the friction angle. Calculate the friction angle based on the relevant forces as determined in Example 3.7.

Solution
By reference to the Merchant's force circle drawn in Example 3.7, the friction angle, $\beta = 50°$.

$F = 1600$ N, $N = 1300$ N, $\beta = ?$

$$\mu = \frac{F}{N} = \frac{1600}{1300} = 1.23 \quad (\text{see Equation 3.6})$$

$$\tan \beta = \mu = 1.23 \quad (\text{see Equation 2.16})$$

$$\beta = \tan^{-1}(1.23) = 50.8°$$

The friction angle β values, as determined by the two methods, are in good agreement.

EXAMPLE 3.9: VERIFICATION OF THE DATA FROM MERCHANT'S FORCE CIRCLE BY USING FORMULAS

a. Verify/compare the measured force values in Example 3.7 by using Equations 3.2–3.5.
b. Verify/compare the measured friction angle by using Merchant's Equation.

Solution
a. By using Equations 3.2–3.5,

$$F = F_c \sin\alpha + F_t \cos\alpha = (1560\sin 10°) + (1270\cos 10°) = 1522\,\text{N}$$

$$N = F_c \cos\alpha - F_t \sin\alpha = (1560\cos 10°) - (1270\sin 10°) = 1316\,\text{N}$$

$$F_s = F_c \cos\phi - F_t \sin\phi = (1560\cos 25°) - (1270\sin 25°) = 877\,\text{N}$$

$$N_s = F_c \sin\phi + F_t \cos\phi = (1560\sin 25°) + (1270\cos 25°) = 1810\,\text{N}$$

The calculated cutting forces are in agreement with the measured forces in Example 3.7.
b. By using the modified form of Merchant's equation (Equation 2.15),

$$\beta = 2\left(45° + \frac{\alpha}{2} - \varphi\right) = 2\left(45° + \frac{10}{2} - 25\right) = 50°$$

$$\beta = \text{friction angle} = 50°$$

The calculated friction angle β is in good agreement with the measured β in Example 3.7.

EXAMPLE 3.10: CALCULATING THE CUTTING POWER

In a turning operation for machining cast iron, the surface speed is 18 m/min, and the cutting force is measured to be 430 N. Calculate the cutting power.

Solution
$F_c = 430$ N, $V = 18$ m/min = 0.3 m/s, $P_c = ?$
By using Equation 3.7,

$$\text{Cutting power} = P_c = F_c V = 430 \times 0.3 = 129\,\text{W}$$

EXAMPLE 3.11: CALCULATING THE GROSS MOTOR POWER FOR MACHINING

The cutting force and the cutting speed in an orthogonal machining operation are 1600 N and 38 m/min, respectively. Calculate the gross motor power for machining if the machine tool motor's efficiency is 90%.

Solution
$F_c = 1600$ N, $V = 38$ m/min = 0.63 m/s, $\eta = 90\% = 0.90$.
By using Equation 3.8,

$$\text{Gross motor power} = \frac{F_{c \cdot v}}{\eta} = \frac{1600 \times 0.63}{0.90} = 1126\,\text{W} = 1.12\,\text{kW}$$

EXAMPLE 3.12: COMPUTING THE MRR WHEN THE CUTTING POWER IS KNOWN

By using the data in Example 3.10 and Table 3.1, compute the minimum metal removal rate for machining of cast iron.

Solution
The maximum unit power for cast iron = $P_u = 5.5$ W·s/mm^3; $P_c = 129$ W, MRR = ?
By using the modified form of Equation 3.9,

$$\text{The minimum metal removal rate} = \text{MRR} = \frac{P_c}{P_u} = \frac{129}{5.5} = 23.45\,\text{mm}^3/\text{s}.$$

EXAMPLE 3.13: CALCULATING THE SPECIFIC ENERGY FOR FRICTION

The cutting force and the thrust force in a straight turning operation are 1400 and 950 N, respectively. The rake angle is 7°. The depth of cut is 0.6 mm, and the thickness of the deformed chip is 0.8 mm. The width of cut is 5 mm. Calculate the specific energy for friction.

Solution

$\alpha = 7°$, $F_c = 1400$ N, $F_t = 950$ N, $t_o = 0.6$ mm, $t_c = 0.8$ mm, $w = 5$ mm, $U_f = ?$
By using Equation 3.2,

$$F = F_c \sin \alpha + F_t \cos \alpha = (1400 \sin 7°) + (950 \cos 7°)$$
$$= 170.6 + 943 = 1113.6 \text{ N}$$

$$r = \frac{t_o}{t_c} = \frac{0.6}{0.8} = 0.75$$

By using Equation 3.13,

$$U_f = \frac{F \cdot r}{wt_o} = \frac{1113.6 \times 0.75}{5 \times 0.6} = 278.4 \frac{\text{N}}{\text{mm}^2} = 278.4 \frac{\text{J/m}}{\text{mm}^2}$$

The specific energy for friction $= U_f = 278.4 \dfrac{J}{1000\,\text{mm} \times \text{mm}^2} = 0.278 \text{ W·s/mm}^3.$

EXAMPLE 3.14: CALCULATING THE SPECIFIC ENERGY FOR SHEARING

By using the data in Example 3.13, calculate the specific energy for shearing, if the shear plane angle is 23°. The shearing speed is 35 m/min, and the cutting speed is 40 m/min.

Solution

$F_c = 1400$ N, $F_t = 950$ N, $\phi = 23°$, $V_s = 25$ m/min $= 0.417$ m/s, $V = 40$ m/min $= 666.67$ mm/s, $t_o = 0.6$ mm, $w = 5$ mm, $U_s = ?$
By using Equation 3.4,

$$F_s = F_c \cos\phi - F_t \sin\phi = (1400\cos 23°) - (950\sin 23°) = 917.5 \text{N}$$

Using Equation 3.14,

$$U_s = \frac{F_s \cdot V_s}{wt_o V} = \frac{917.5 \times 0.417}{5 \times 0.6 \times 666.67} = 0.191 \text{N/mm}^2$$

The specific energy for shearing $= U_s = 0.191$ W·s/mm³.

EXAMPLE 3.15: CALCULATING THE TOTAL SPECIFIC ENERGY IN MACHINING

By using the data in Examples 3.13 and 3.14, calculate the total specific energy in machining; and hence identify the metal that was machined.

Solution

By using Equation 3.15,

The total specific energy in machining $= U = U_f + U_s$
$$= 0.278 + 0.191 = 0.47 \text{W} \cdot \text{s/mm}^3$$

By reference to the data in Table 3.1, the metal that was machined is either an aluminum alloy or a magnesium alloy.

EXAMPLE 3.16: CALCULATING THE PERCENTAGE OF FRICTION ENERGY IN MACHINING

Derive an expression for the percentage of the specific energy for friction in machining, and hence calculate the percentage by using the data in Examples 3.13–3.15.

Solution

$$\% \text{ Specific energy for friction} = \frac{\text{Speciifc energy for friction}}{\text{Total specific energy}} \times 100$$

$$= \frac{U_f}{P_u} \times 100 = \frac{\dfrac{F \cdot r}{w\, t_o}}{\dfrac{F_c}{w\, t_o}} \times 100 = \frac{F \cdot r}{F_c} \times 100$$

By reference to Example 3.13, $F_c = 1400$ N, $F = 1113.6$ N, and $r = 0.75$

$$\% \text{ Specific energy for friction} = \frac{F \cdot r}{F_c} \times 100 = \frac{1113.6 \times 0.75}{1400} \times 100 = 59.6$$

EXAMPLE 3.17: CALCULATING THE MEAN TEMPERATURE IN MACHINING

An orthogonal machining operation was performed at a cutting speed of 2 m/s with the depth of cut = 0.013 cm. The work material properties are as follows: flow stress = 120 MPa, thermal diffusivity = 97 mm²/s, and volumetric specific heat = 484 cm·kg/cm³ °C Calculate the mean temperature at the tool–chip interface.

Solution
$Y_f = 120$ MPa = 120,000 kPa, $V = 2$ m/s = 200 cm/s, $t_o = 0.013$ cm, $c = 384$ cm·kg/cm³·°C, $K = 97$ mm²/s = 97 × (0.1 cm)²/s = 0.97 cm²/s
By using Equation 3.16,

$$T = \frac{1.2 Y_f}{c} \left(\frac{V t_o}{K} \right)^{0.33} = \frac{1.2 \times 120,000}{384} \left(\frac{200 \times 0.013}{0.97} \right)^{0.33} = 518\,K$$

The mean temperature at the tool–chip interface = 518 K = 518–273 = 245°C.

QUESTIONS AND PROBLEMS

3.1 Encircle the most appropriate answers for the following statements.
   a. The cutting force needed to cut a chip area of 1 mm² that has a thickness of 1 mm is called:
      (i) the thrust force, (ii) the specific force, (iii) the friction force, and (iv) the normal force

b. Which force, if too high, will result in pushing away of the cutting tool from workpiece?
(i) the thrust force, (ii) the specific force, (iii) the shear force, (iv) the cutting force.
c. Which force acts on the workpiece and supplies the energy required for machining?
(i) the thrust force, (ii) the normal force, (iii) the friction force, (iv) the cutting force.
d. In straight turning, the thrust force is called:
(i) the friction force, (ii) the normal force, (iii) the feed force, (iv) the cutting force.
e. Which force acts along the tool–chip interface?
(i) the friction force, (ii) the normal force, (iii) the feed force, (iv) the shear force.
f. Which forces can be measured by a *force dynamometer*?
(i) friction and shear forces, (ii) cutting and normal forces, (iii) cutting and thrust forces.
g. The cutting power per unit volume rate of metal removed is called:
(i) specific power, (ii) gross motor power, (iii) specific energy, (d) cutting power
h. The cutting power divided by the machine tool motor efficiency is called:
(i) specific power, (ii) gross motor power, (iii) specific energy, (d) cutting power.

3.2 Diagrammatically illustrate the forces acting on the workpiece, tool, and the chip during orthogonal machining. Compare the forces with those acting on the work during straight turning.

3.3 Explain the working principle of a *force dynamometer*.

P3.4 Refer to the graphical plot [Figure P3.4] obtained by loading/unloading by using a force dynamometer oriented for measuring the thrust force F_t.
a. By using the data in Figure P3.4, obtain the calibration curve, and hence calculate the constants a and b for use in Equation 3.1.
b. The output of DAQ during turning operation, with a depth of cut, DOC of 3 mm, shows a strain of 6×10^{-5}. Calculate the thrust force for the specified DOC.

P3.5 The cutting force and the thrust force, in an orthogonal cutting operation, are 1300 and 1000 N, respectively. The rake angle is 8°, and the shear plane angle is 26°. Calculate (a) the friction force, (b) the normal force, (c) the shear force, and (d) the normal to shear force.

P3.6 By using the data in P3.5, determine the friction force, normal force, shear force, and the normal to shear force by drawing the Merchant's Force Circle. Compare your results.

P3.7 By using the data in P3.5–P3.6, determine the friction angle by three methods.

P3.8 The cutting force and the cutting speed in an orthogonal machining operation are 1500 N and 33 m/min, respectively. Calculate (a) the gross motor power for

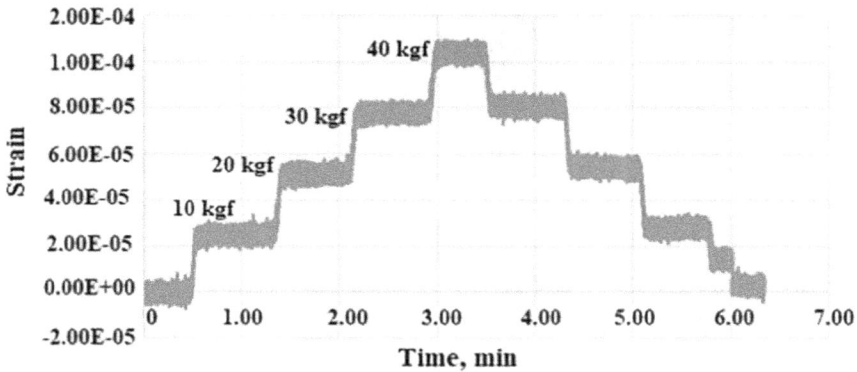

FIGURE P3.4 Strain–time plot for determining the constants *a* and *b* for measuring the thrust force

machining if the machine tool motor's efficiency is 87%, (b) the minimum metal removal rate for machining of aluminum.

P3.9 Derive an expression for the percentage of the specific energy for shear in machining, and hence calculate the percentage by using the data in Examples 3.13–3.15. Verify that total of the % friction energy and the % shear energy equals 100.

REFERENCES

Astakhov, V.P. (1998), *Metal Cutting Mechanics*, CRC Press, Boca Raton, FL.

Cica, D., Sredanovic, B., Lakic-Globocki, G., Kramar, D. (2015), Modeling of the cutting forces, in turning process using various methods of cooling and lubricating: An artificial intelligence approach, *Advances in Mechanical Engineering*, 5, 798597.

Groover, M.P. (2016), *Groover's Principles of Modern Manufacturing: Materials, Processes, and Systems, SI Version*, 6th Edition, John Wiley & Sons Inc., New York.

Huda, Z. (2018), *Manufacturing: Mathematical Models, Problems, and Solutions*, CRC Press, Boca Raton, FL.

Kesavan, R., Ramnath, B.V. (2010), *Machine Tools*, University Science Press, New Delhi.

Yaldız, S., Ünsaçar, F. (2006), A dynamometer design for measurement the cutting forces on turning, *Measurement*, 39(1), 80–89.

4 Cutting Tools and Materials

4.1 CUTTING TOOLS AND MATERIALS: *FUNDAMENTALS*

4.1.1 CUTTING TOOLS AND THEIR TYPES

A cutting tool (or a cutter) is a wedge-shaped and sharp-edged tool that is used to remove material from a workpiece by means of shear deformation in a conventional machining process. In machining, the machine tool holds the cutter and workpiece and at the same time its mechanism provides the desired cutting conditions (speed, feed, and depth of cut). The shape and size of a cutter may vary depending on the feature to be produced and the operation employed.

A *cutting edge* is a straight or curved edge produced by intersection of any two tool point surfaces (*e.g.*, intersection of rake face and flank surface) (see Figure 4.1). Based on the number of cutting edge(s) present, cutting tools may be divided into two types: (a) single-point and (b) multi-point cutters. A *single-point cutting tool* contains just one main cutting edge, which removes entire volume of material in a single pass (see Figure 4.1a). Examples of single-point cutting tools include the cutters used in turning, shaping, planing, boring, and the like. A *multi-point cutting tool* has at least two cutting edges; all of these cutting edges can equally participate in material removal in a single pass (see Figure 4.1b). Milling, grinding, drilling, reaming, hobbing, etc. operations involve the use of multi-point cutters.

4.1.2 CUTTING TOOL MATERIALS

The performance of a cutting tool strongly depends on its material's properties. The material of a cutting tool must be harder than that of the work to be machined.

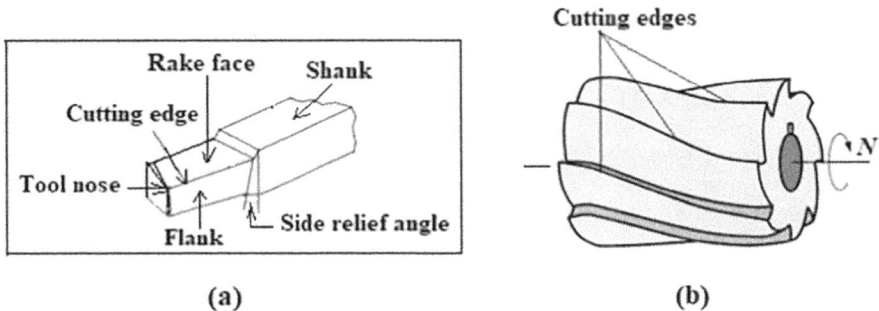

FIGURE 4.1 Cutting tools (a) single-point cutting tool and (b) multiple-point (milling) cutter

TABLE 4.1
Cutting Tool Materials and Their Hot-Hardness Temperatures

Cutting tool material	High-carbon steel	High-speed steel	Cemented carbide	Coated carbides	Ceramics	Cubic boron nitride (CBN)	Synthetic diamond
Hot-hardness temperature	200°C	650°C	900–1000°C	1100°C	1200°C	1500°C	1500°C

Besides hardness, the cutting tool material must have a high resistance to wear. Additionally, the cutting tool material must have a good hot hardness *i.e.*, the hardness and strength of the cutting tool must be maintained at high temperatures. Another important requirement of a cutting tool material is its toughness so as to avoid brittle fracture, especially during interrupted cutting operations.

Commonly used cutting tool materials include high carbon steel, high-speed steel (HSS), cemented carbide, coated carbide, ceramic, cubic boron nitride (CBN), and synthetic diamond. The performance of a cutting tool material is mainly dictated by its *hot-hardness temperature* – the temperature up to which the hot hardness is maintained. The material with a high hot-hardness temperature enables a machinist to perform machining operations at reasonably high cutting speeds. The hot-hardness temperatures of various cutting tool materials are listed in Table 4.1 (Davis, 1995).

4.2 TOOL WEAR AND FAILURE

It has been explained in the preceding chapter that the cutting forces and temperature have pronounced effects on the cutting tool life. In particular, a significant rise in cutting temperature is detrimental to the cutting tool since a high temperature may result in tool failure (Kalpakjian, 1982). A cutting tool is said to have failed if one or more of the following conditions are reached: (1) fracture or deformation of the tool occurs, (2) the cutting force increases to a set limit, (3) the surface finish becomes unacceptable, (4) the machined workpiece count exceeds a set limit, and (5) the flank wear band (VB) reaches a limit; for HSS tools, $VB_{max} = 1.5$ mm, while for carbide tools $VB_{max} = 0.4$ mm (see Figure 4.2).

A tool may fail by one of the following modes of failure: (1) fracture failure, (2) temperature failure, or (3) gradual wear. The *fracture failure* occurs when a crack initiates and propagates to fast fracture (sudden failure) or when the cutting force exceeds a set limit. It can be avoided by ensuring adequate fracture toughness in the tool material and/or by reducing cutting forces. The *temperature failure* refers to the deformation of the tool point and loss of sharp cutting edge, which occurs when the cutting temperature becomes too high for the tool material thereby causing softening of the tool point. The temperature failure can be avoided by using a tool material with high hot-hardness and by using an effective cutting fluid – coolant. The *gradual wear* occurs gradually at rake face (as crater wear), at flank (*flank wear*), or at notch (*notch wear*) (see Figure 4.2). It can be minimized by controlling the friction force

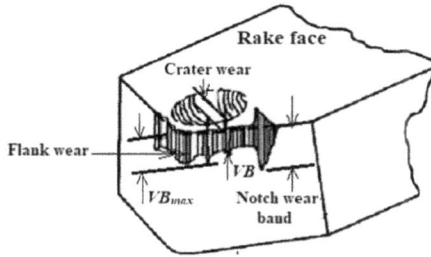

FIGURE 4.2   Gradual wear of cutting tool. (VB = flank wear band)

that acts on the tool–chip interface and by reducing the cutting speed (Drouillet, et al., 2016).

4.3 CUTTING TOOL LIFE – *TAYLOR'S TOOL LIFE EQUATION*

Tool life is the service time to total failure of the cutting tool at a certain cutting speed. The life of a cutting tool depends on several factors, including the cutting speed, temperature, cutting forces, and the like. The main factor determining a tool's life is the cutting speed. The relationship between tool life and cutting speed is expressed as the *Taylor's tool life equation*, as follows (Taylor, 1907):

$$V T_L^n = C \qquad (4.1)$$

where V is the cutting speed, m/min; T_L is the tool life, min; and n and C are constants. The n values usually range from 0.1 (for carbon steel tools) to 0.7 (for ceramic tools) (see Table 4.2). The value of C depends whether the workpiece material is steel or non-steel (see Table 4.2).

By taking logarithm of both sides of Equation 4.1, we obtain:

$$\log V = -n \log T_L + \log C \qquad (4.2)$$

TABLE 4.2
The n and C Values for Commonly Used Cutting Tool Materials

Cutting Tool	High Carbon Steel	High Speed Steel	Cemented Carbide	Coated Carbide	Ceramic
n	0.1	0.125	0.25	0.25	0.7
C (for steel cutting)	20 m/min	70 m/min	500 m/min	700	3000
C (for non-steel cutting)	70 m/min	120 m/min	900 m/min	–	–

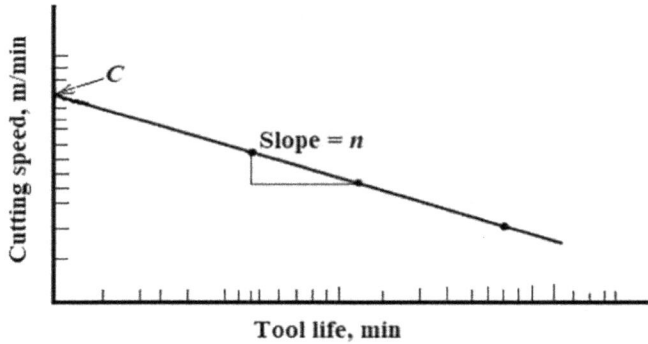

FIGURE 4.3   The graphical representation (log-log scale) of Taylor's tool-life equation

Equation 4.2 is a linear equation in slope-intercept form; it can be graphically represented as shown in Figure 4.3.

The slope of the curve in Figure 4.3 represents the exponent n in the Taylor's tool life equation.

$$n = \text{slope} = \frac{\log V_2 - \log V_1}{\log T_{L2} - \log T_{L1}} \tag{4.3}$$

It is evident in Equation 4.3 that a high n value of a cutting tool indicates its good performance since it enables a machinist to perform the machining operation at higher cutting speeds. Additionally, Figure 4.3 enables us to define the constant C in the Taylor's tool life equation as follows: C is the cutting speed for which the tool life is equal to 1 min. The significance of Equations 4.1–4.3 is illustrated in Examples 4.1–4.14.

Although the main machining parameter that influences the tool life is the cutting speed, other cutting conditions also play roles in determining the tool life. The tool life is related to the cutting speed, the feed, and the depth of cut for a turning operation as follows (Sharma, 2008):

$$V\,T_L^{0.13} f^{0.77} d^{0.37} = C \tag{4.4}$$

where T_L is the tool life, min; C is a constant, V is the cutting speed, m/min; d is the depth of cut, mm; and f is the feed, mm/rev (see Example 4.15).

4.4 CUTTING FLUIDS

Cutting fluids are the fluids that are specifically designed for use as either a coolant or lubricant (or both) in machining operations. They are used in machining operations for a variety of reasons, including (a) to cool the tool and work, (b) to reduce friction between tool and chip interface resulting in easier machining, (c) to avoid temperature failure of the tool, (d) to improve tool life, (e) to reduce workpiece thermal deformation, (f) to improve surface finish due to lubrication, and (g) to flush away chips from the cutting zone.

A good cutting fluid must have the following eight characteristics: (1) good cooling capability, (2) good lubricating ability, (3) corrosion resistance, (4) relatively low viscosity, (5) long shelf life, (6) transparency, and (7) non-toxicity, and (8) non-flammability. Based on the type of functions, cutting fluids may be broadly divided into two types: (a) coolants and (b) lubricants. *Coolants* are designed to reduce the effects of heat in machining, whereas lubricants are designed to reduce tool-chip and tool-work friction. *Coolants* are generally water-base cutting fluids; they are very effective at high cutting speeds where high temperatures are problems. When performing a machining operation using a tool material with a lower hot hardness (e.g., HSS tool), it is important to recommend a *coolant* as the cutting fluid. *Lubricants* are usually oil-based cutting fluids; they are most effective in reducing tool-chip friction at lower cutting speeds, thereby improving tool life and producing better surface finish of the machined surface. Besides reducing friction, *lubricants* also reduce temperature in the machining operation (Youssef and El Hofy, 2008).

4.5 CALCULATIONS – *WORKED EXAMPLES ON CUTTING TOOLS AND MATERIALS*

EXAMPLE 4.1: GRAPHICAL CALCULATION OF CONSTANTS N AND C IN TAYLOR'S TOOL LIFE EQUATION

Machining experiments at various cutting speeds V were performed, and the corresponding tool lives T_L were recorded as presented in Table E-4.1. (a) Plot a graph using the data in Table E-4.1. (b) Determine the constants n and C from the plot drawn by you in (a).

Solution
The cutting speed-tool life graphical plot on the log-log scale is shown in Figure E-4.1.
  From the graphical plot in Figure E-4.1, $C = 300$ m/min.
  By using Equation 4.3 for the two data points shown in Figure E-4.1,

$$n = \text{slope} = \frac{130-160}{12-5} = -4.28$$

The constants are $n = 4.28$ and $C = 300$ m/min.

TABLE E-4.1
Cutting Speed-Tool Life Data

V (m/min)	200	180	160	130	100
TL (min)	2	3	5	12	41

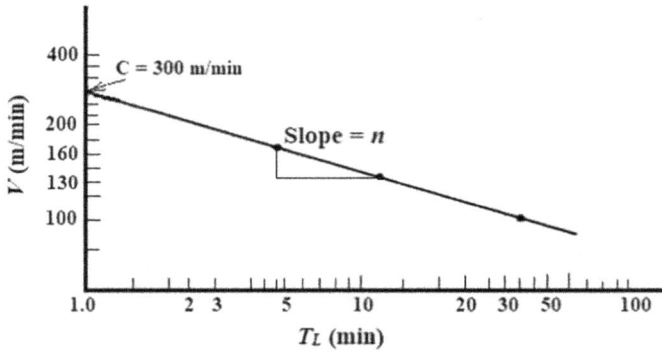

FIGURE E-4.1   Cutting speed-tool life plot using log-log scale

EXAMPLE 4.2: CALCULATING THE CUTTING SPEED FOR A SPECIFIED CUTTING TOOL AND ITS LIFE

It is desired to achieve a tool life of 300 min for a ceramic tool for a turning operation to machine a steel workpiece. What cutting speed do you recommend for the tool? Refer to Table 4.2.

Solution

By reference to Table 4.2, for ceramic tool for steel cutting, $n = 0.7$, and $C = 3000$ m/min.

By using Equation 4.1,

$$V(300)^{0.7} = 3000$$

$$V = \frac{3000}{54.2} = 55.35 \, \text{m/min}$$

The recommended cutting speed = 55 m/min.

EXAMPLE 4.3: CALCULATING THE PERCENT CHANGE IN THE CUTTING SPEED FOR TWO TOOLS

It is desired to achieve a tool life of 300 min for a HSS tool for a turning operation to machine a steel workpiece. What cutting speed do you recommend for the tool? Compare your result with the use of ceramic tool and calculate the percent decrease in the cutting speed.

Solution

By reference to Table 4.2, for HSS tool for steel cutting, $n = 0.125$ and $C = 70$ m/min.

By using Equation 4.1,

$$VT_L^n = C$$

$$V(300)^{0.125} = 70$$

$$V = \frac{70}{2.04} = 34.3\,\text{m/min}$$

By changing from ceramic tool to HSS tool, the % decrease in cutting speed is:

$$\% \text{ Decrease in cutting speed} = \frac{V_1 - V_2}{V_1} \times 100 = \frac{55 - 34}{55} \times 100 = 38$$

EXAMPLE 4.4: THE QUANTITATIVE EFFECT OF TOOL CHANGE ON MATERIAL REMOVAL RATE (MRR)

By using the data in Examples 4.2 and 4.3, calculate the MRR for the following tools if a feed of 0.3 mm/rev. and depth of cut of 2.2 mm were used for the turning operation: (a) ceramic tool and (b) HSS tool.

Solution
It is learnt in Chapter 3 (Section 3.2.1) that in straight turning, the uncut chip thickness = depth of cut = t_o and the width of cut = feed = f. Thus Equation 3.10 can be modified as follows:

$$\text{MRR} = V\,d\,f \qquad\qquad (\text{E-4.4})$$

a. For ceramic tool, $V = 55.35$ m/min = 55,350 mm/min, $d = 2.2$ mm, $f = 0.3$ mm/rev
   By using Equation E-4.4 for ceramic tool,

$$\text{MRR} = V\,d\,f = 55,350 \times 2.2 \times 0.3 = 36,531\,\text{mm}^3/\text{min}$$

b. By using Equation E-4.4 for HSS tool,

$$\text{MRR} = V\,d\,f = 34,300 \times 2.2 \times 0.3 = 22,638\,\text{mm}^3/\text{min}$$

EXAMPLE 4.5: CALCULATING THE % CHANGE IN THE CUTTING SPEED FOR TWO WORK MATERIALS

It is desired to achieve a tool life of 300 min for a HSS tool. What cutting speed do you recommend for cutting a non-steel work using the tool? Compare your result with the steel cutting (Example 4.3) and calculate the percent increase in the cutting speed.

Solution
By reference to Table 4.2, for non-steel cutting using HSS tool, $n = 0.125$ and $C = 120$ m/min.
   By using Equation 4.1,

$$V(300)^{0.125} = 120$$

$$V = \frac{120}{2.04} = 58.8 \, \text{m/min}$$

By changing from steel cutting to non-steel cutting, the % increase in the cutting speed is:

$$\% \text{ Increase in cutting speed} = \frac{58.8 - 34.3}{34.3} \times 100 = 71.4$$

EXAMPLE 4.6: THE EFFECT OF CUTTING SPEED ON TOOL LIFE

The tool life of a HSS tool ($n = 0.125$) at a cutting speed of 20 m/min is 2.8 h. Calculate the tool life when the machining operation is performed at 25 m/min.

Solution
For $V = 20$ m/min, $T_L = 2.8$ h $= 168$ min, $n = 0.125$

$$VT_L^n = C \quad (\text{Equation } 4.1)$$

$$20(168)^{0.125} = C$$

$$C = 38$$

By using Equation 4.1 at $V = 25$ m/min,

$$25T_L^{0.125} = 38$$

$$T_L^{0.125} = 1.518$$

$$T_L = 1.518^{\frac{1}{0.125}} = 1.518^8 = 28 \, \text{min}$$

The tool life at cutting speed of 25 m/min is 28 min.

EXAMPLE 4.7: PLOTTING FLANK WEAR AS A FUNCTION OF TIME FOR TWO CUTTING SPEEDS

A series of turning machining tests (on a particular work material by using a specified tool) were performed, and the resulting flank wear data are presented in Table E-4.7. The last wear data value in each row indicates the final tool failure.

TABLE E-4.7
Flank Wear at Various Cutting Times at Specified Cutting Speeds

Cutting time (min)	5	10	15	20	30	40
Flank wear (mm) at $V = 100$ m/min	0.10	0.14	0.20	0.26	0.43	0.47
Flank wear (mm) at $V = 130$ m/min	0.23	0.40	0.57			

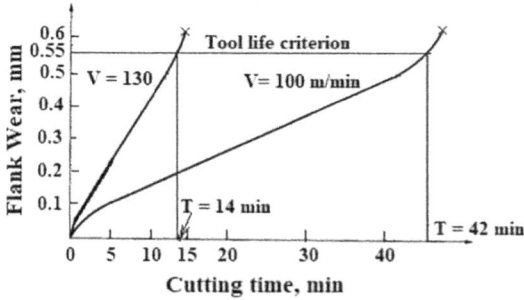

FIGURE E-4.7 The flank wear versus cutting time plot at specified cutting speeds

Plot flank wear as a function of cutting time on a single piece of linear graph paper. Using 0.55 mm of flank wear as the criterion of tool failure, compute tool lives for the two cutting speeds.

Solution
The flank wear versus cutting time graphical plot is shown in Figure E-4.7.
It is evident in Figure E-4.7 that at $V = 100$ m/min, the tool life = 42 min and at $V = 130$ m/min, the tool life = 14 min.

EXAMPLE 4.8: CALCULATING THE CONSTANTS *N* AND *C* IN TAYLOR'S TOOL-LIFE EQUATION

By using the data in Example 4.7, calculate the values of n and C in the Taylor equation.

Solution
By using Equation 4.1 for tool life of 42 min at $V = 100$ m/min, and tool life 14 at $V = 130$,

$$100(42)^n = C \qquad \text{(E4.7-a)}$$

$$130(14)^n = C \qquad \text{(E4.7-b)}$$

By combining Equations E4.7-a and E4.7-b,

$$100(42)^n = 130(14)^n$$

$$n = 0.239$$

By substituting $n = 0.239$ in Equation E4.7-a,

$$100(42)^{0.239} = C$$

$$C = 100 \times 2.44 = 244$$

Thus the constants are $n = 0.239$ and $C = 244$.

EXAMPLE 4.9: COMPUTING THE TOOL LIFE USING THE FLANK WEAR DATA

By using the data in Example 4.8, calculate the tool life at a cutting speed of 85 m/min.

Solution
$n = 0.239$, $C = 244$, $V = 85$ m/min, $T_L = ?$

$$VT_L^n = C \quad (\text{Equation 4.1})$$

$$85T_L^{0.239} = 244$$

$$T_L^{0.239} = 2.87$$

$$T_L = 2.87^{\frac{1}{0.239}} = 2.87^{4.18} = 82\,\text{min}$$

EXAMPLE 4.10: COMPUTING THE CUTTING SPEED USING THE FLANK WEAR DATA

By using the data in Example 4.8, calculate the cutting speed for a tool life of 30 min.

Solution
$n = 0.239$, $C = 244$, $T_L = 30$ min, $V = ?$

$$VT_L^n = C \quad (\text{Equation 4.1})$$

$$V(30)^{0.239} = 244$$

$$2.254\,V = 244$$

$$V = 108.25\,\text{m/min}$$

The required cutting speed = 108.25 m/min.

EXAMPLE 4.11: CALCULATING THE MRRs AT TWO CUTTING SPEEDS

By using the data in Example 4.7, calculate the MRR for the two cutting speeds, if the feed was 0.28 mm/rev and the depth of cut was 3.0 mm.

Solution
By using Equation E-4.4 at cutting speeds 100 m/min and 130 m/min,

$$\text{MRR}_{(100)} = V\,d\,f = 100{,}000\,\text{mm/min} \times 3\,\text{mm} \times 0.28\,\text{mm/rev} = 84{,}000\,\text{mm}^3/\text{min}$$

$$\text{MRR}_{(130)} = V\,d\,f = 130{,}000\,\text{mm/min} \times 3\,\text{mm} \times 0.28\,\text{mm/rev} = 109{,}200\,\text{mm}^3/\text{min}$$

EXAMPLE 4.12: CALCULATING THE PERCENT DECREASE IN TOOL LIFE BY DOUBLING CUTTING SPEED

What is the percent decrease in tool life of a cemented carbide tool if the cutting speed is doubled?

Solution
Let at the cutting speed V_1 the tool life $= T_{L1}$, and at cutting speed V_2 the tool life $= T_{L2}$

By using Equation 4.1 at the cutting speed V_1,

$$V_1 T_{L1}^n = C \tag{E4.11-a}$$

By using Equation 4.1 at the cutting speed V_2,

$$V_2 T_{L2}^n = C \tag{E4.11-b}$$

By combining Equations E4.11-a and E4.11-b, we obtain:

$$V_1 T_{L1}^n = V_2 T_{L2}^n$$

$$V_1 T_{L1}^n = 2 V_1 T_{L2}^n \quad \left(\text{It is given that } V_2 = 2V_1\right)$$

$$T_{L1}^n = 2 T_{L2}^n$$

$$\left(\frac{T_{L1}}{T_{L2}}\right)^n = 2$$

$$\left(\frac{T_{L1}}{T_{L2}}\right)^{0.25} = 2 \quad \left(\text{for the carbide tool, } n = 0.25\right)$$

$$\frac{T_{L1}}{T_{L2}} = 2^{\frac{1}{0.25}} = 2^4 = 16$$

$$T_{L2} = \frac{T_{L1}}{16} = 0.0625\,T_{L1}$$

$$\% \text{ Decrease in tool life} = \frac{T_{L1} - T_{L2}}{T_{L1}} \times 100 = \frac{T_{L1} - 0.0625\,T_{L1}}{T_{L1}} \times 100 = 94$$

or Decrease in tool life = 94%.

EXAMPLE 4.13: CALCULATING NEW TOOL LIFE WHEN THE CUTTING SPEED IS INCREASED BY 30%

Machining of a mild steel workpiece at a cutting speed of 55 m/min by using a cemented carbide tool resulted in a tool life of 97 min. Calculate the tool life when the cutting speed is increased by 30%.

Solution

$T_{L1} = 97$ min, $V_1 = 55$ m/min, For carbide tool, $n = 0.25$

For 30% higher cutting speed, $V_2 = V_1 + 30\% V_1 = 1.3 V_1$

$$V_1 T_{L1}^n = V_2 T_{L2}^n \quad \left(\text{see Example 4.12} \right)$$

$$V_1 T_{L1}^n = 1.3 V_1 T_{L2}^n$$

$$\left(\frac{T_{L1}}{T_{L2}} \right)^n = 1.3$$

$$\left(\frac{T_{L1}}{T_{L2}} \right)^{0.25} = 1.3$$

$$\frac{T_{L1}}{T_{L2}} = 1.3^{\frac{1}{0.25}} = 1.3^4 = 2.856$$

$$T_{L2} = \frac{T_{L1}}{2.856} = \frac{97}{2.856} = 33.96 \cong 34 \, \text{min}$$

The new tool life = 34 min.

EXAMPLE 4.14: DEVELOPING A TAYLOR'S TOOL LIFE EQUATION FOR SPECIFIED TOOL AND WORK

Machining of a carbon steel workpiece at a cutting speed of 95 m/min by using a high-carbon high-tungsten steel tool ($n = 0.09$) resulted in a tool life of 48 min. Develop the general-specific *Taylor's tool life equation* for the specified tool and the work material.

Solution

By using modified form of Equation 4.2,

$$\log C = \log V + n \log T_L$$

$$\log C = \log 95 + 0.09 \log 48$$

$$C = 134$$

By substituting $n = 0.09$ and $C = 134$ in the Taylor's tool life equation (Equation 4.1), we obtain the general-specific Taylor's tool life equation as follows:

$$VT_L^{0.09} = 134$$

EXAMPLE 4.15: COMPUTING NEW TOOL LIFE WHEN THE THREE CUTTING CONDITIONS CHANGE

An 80-min tool life was obtained in a machining operation at a cutting speed of 28 m/min, feed of 0.3 mm/rev, and depth of cut = 2.3 mm. Calculate the tool life if the three machining parameters are increased by 25%.

Solution

$T_L = 80$ min, $V = 28$ m/min, $f = 0.3$ mm/rev, $d = 2.3$ mm, $C = ?$

By using Equation 4.4,

$$C = V T_L^{0.13} f^{0.77} d^{0.37} = 28 \times (80)^{0.13} (0.3)^{0.77} (2.3)^{0.37}$$
$$= 28 \times 1.76 \times 0.395 \times 1.36 = 26.47$$

New cutting speed = $V' = 28 + (25\% \times 28) = 35$ m/min.
New feed = $f' = 0.3 + 25\% \times 0.3 = 0.375$ mm/rev.
New depth of cut = $d' = 2.3 + 25\% \times 2.3 = 2.875$ mm.
By using Equation 4.4 for the new machining parameters,

$$V T_L^{0.13} f^{0.77} d^{0.37} = C$$

$$(35) T_L^{0.13} (0.375)^{0.77} (2.875)^{0.37} = 26.47$$

$$T_L^{0.13} = \frac{26.47}{35 \times 0.47 \times 1.48} = 1.087$$

$$T_L = 1.087^{\frac{1}{0.13}} = 1.087^{7.69} = 1.9 \, \text{min}$$

The new tool life = 1.9 min.

QUESTIONS AND PROBLEMS

4.1 Encircle the most appropriate answers.
  a. Which cutting tool material has the highest hot hardness temperature?
     (i) cemented carbide, (ii) CBN, (iii) HSS, (iv) high carbon steel
  b. Which cutting tool material has the lowest hot hardness temperature?
     (i) cemented carbide, (ii) CBN, (iii) HSS, (iv) high carbon steel
  c. What is the maximum allowable limit of flank wear band for carbide tools?
     (i) 0.4 mm, (ii) 1.0 mm, (iii) 1.5 mm, (iv) 2.0 mm
  d. What is the maximum allowable limit of flank wear band for HSS tools?
     (i) 0.4 mm, (ii) 1.0 mm, (iii) 1.5 mm, (iv) 2.0 mm
  e. Which type of failure can be avoided by ensuring high toughness in the tool material?
     (i) fracture failure, (ii) temperature failure, (iii) gradual wear, (iv) all types of failure
  f. Which type of failure can be avoided by minimizing the friction force during machining?

(i) fracture failure, (ii) temperature failure, (iii) gradual wear, (iv) all types of failure

g. Which are the two main functions of a cutting fluid in machining?
(i) to improve surface finish and tool life, (ii) to reduce forces and power, (iii) to reduce temperature and friction at the tool–chip interface, (iv) remove heat and wash away chips

h. Which cutting condition has the most pronounced effect on tool life?
(i) cutting speed, (ii) feed, (iii) depth of cut, (iv) equal effect of all cutting conditions.

4.2 Differentiate between single-point and multiple-point cutting tools and give examples.

4.3 a. List the properties required in a cutting tool material and explain them.
b. List the various cutting tool materials mentioning their hot-hardness temperatures.

4.4 a. Diagrammatically illustrate the gradual wear in a single-point cutting tool.
b. What are the criteria for a cutting tool failure?

4.5 What are the various modes of failure of a cutting tool? How can they be avoided?

4.6 a. Why are cutting fluids used in machining?
b. List the characteristics of a cutting fluid.
c. Differentiate between coolants and lubricants.

P4.7 The tool life of a ceramic tool at a cutting speed of 80 m/min is 200 min. Calculate the tool life when the machining operation is performed at 60 m/min.

P4.8 A series of turning machining tests (on a specified work material and tool) were performed, and the resulting flank wear data are presented in Table P4.8. The feed was 0.3 mm/rev, and the depth of cut was 3.5 mm. The last wear data value in each row indicates the final tool failure.

a. Plot flank wear as a function of cutting time on a single piece of linear graph paper.
b. Using 0.55 mm of flank wear as the criterion of tool failure, determine the tool lives for the two cutting speeds using the graphical plot.
c. Calculate the values of n and C in the Taylor equation.
d. Compute the MRR for the two cutting speeds.

P4.9 What is the percent decrease in tool life of a HSS tool if the cutting speed is tripled?

TABLE P4.8
Cutting Time-Flank Wear Data

Cutting time (min)	2	3	4	5	6	10	15
Flank wear (mm) at $V = 130$ m/min	0.13	0.18	0.21	0.23	0.28	0.4	0.57
Flank wear (mm) at $V = 160$ m/min	0.14	0.28	0.47	0.50	0.57		

P4.10 Machining of a mild steel workpiece at a cutting speed of 50 m/min by using a ceramic tool resulted in a tool life of 160 min. Calculate the tool life when the cutting speed is increased by 20%.

P4.11 Machining of a steel workpiece at a cutting speed of 85 m/min by using a high-carbon steel resulted in a tool life of 20 min. Develop the general-specific *Taylor's tool life equation* for the specified tool and the work material.

P4.12 A 70 min tool life was obtained in a machining operation at a cutting speed of 32 m/min, feed of 0.3 mm/rev, and depth of cut = 2.5 mm. Calculate the tool life if the three machining parameters are increased by 30%.

REFERENCES

Davis, J.R. (ed.) (1995), *Tool Materials*, ASM International, Metals Park, OH.

Drouillet, C., Karandikar, J., Nath, C., Journeaux, A.-C., El Mansori, M., Kurfees, T. (2016), Tool life prediction in milling using spindle power with the neural network technique, *Journal of Manufacturing Processes*, 22, 161–168.

Kalpakjian, S. (ed.) (1982), *Tool and Die Failures*, ASM International, Metals Park, OH.

Sharma, P.C. (2008), *A Textbook of Manufacturing Technology-II*, S. Chand and Company (Pvt) Ltd., New Delhi.

Taylor, F.W. (1907), *On the Art of Cutting Metals*, Transactions of American Society of Mechanical Engineers (Trans. ASME, 28), New York.

Youssef, H.A., El Hofy, H. (2008), *Machining Technology: Machine Tools and Operations*, CRC Press, Boca Raton, FL.

Part II

Conventional Machining and Gear Manufacturing

5 Turning Operations and Machines

5.1 TURNING OPERATIONS AND THEIR INDUSTRIAL APPLICATIONS

Turning is a machining operation that involves the removal of material from the outer diameter of a rotating cylindrical workpiece by using a single-point cutting tool (see Figure 1.1). In *turning*, the speed motion is provided by the rotational speed of the workpiece, and the feed motion is achieved by moving the cutting tool slowly in a direction parallel to the axis of rotation of the workpiece. *Turning* operation is performed by use of a *lathe* – turning machine; here the work is attached to a spindle, whereas the tool is clamped in the tool post of the lathe (see Figure 5.1). During *turning*, the workpiece is held firmly in place by either one or two rigid supports; the latter are called *centers*. In general, the workpiece is secured about its axis of rotation using a chuck or face plate (see Figure 5.1). In manufacturing, it is significant to create cylindrical workpieces as per specifications; this goal can be economically accomplished by using a *lathe*. Although the *lathe* enables us to produce a symmetrical object out of wood, metal, plastic, etc. (Yates, 1922), it is best considered for the machining works on metallic materials.

The *lathe* is one of the basic machine tools for industrial applications since a wide variety of shapes and forms can be created by turning and related operations. In industrial practice, *lathes* produce a wide range of machined parts; these include drive shafts of cars, gun barrel, crankshaft, screws, locomotive wheels, legs for

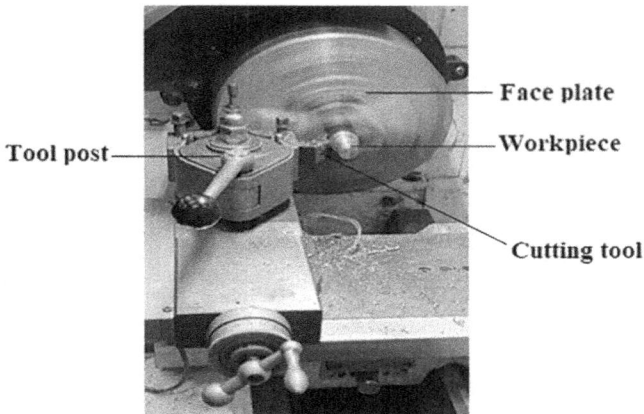

FIGURE 5.1 Turning operation on a *lathe*

tables, and many other engineering components. Heavy-duty large-scale lathes can
be used to machine a giant metal cone or disc, while small-scale machines can cut out
even a chess piece. A *lathe* revolves the workpiece on its axis for performing a wide
range of machining operations; these turning-related operations include straight turning, taper turning, cutting, knurling, boring, reaming, grooving, chamfering, facing,
threading, and the like. There are various types of *lathe machines* to meet the diversified industrial needs (see Section 5.3).

5.2 TURNING-RELATED OPERATIONS

In industrial practice, there are often machining requirements to produce cylindrical
parts with complex geometric features that require multiple turning-related operations. Thus, during the machining process cycle, a variety of turning-related operations may be performed to the workpiece to yield the desired part of shape. These
turning-related operations include straight turning, taper turning, facing, boring, contour turning, form turning, cut-off, grooving, external threading, internal threading,
knurling, and the like (see Figure 5.2). External turning operations modify the outer
diameter of the workpiece, while internal turning operations modify the inner diameter. Each type of turning operation is defined by the type of cutter used and the path
of the cutter to remove material from the workpiece, as explained in the following
paragraph.

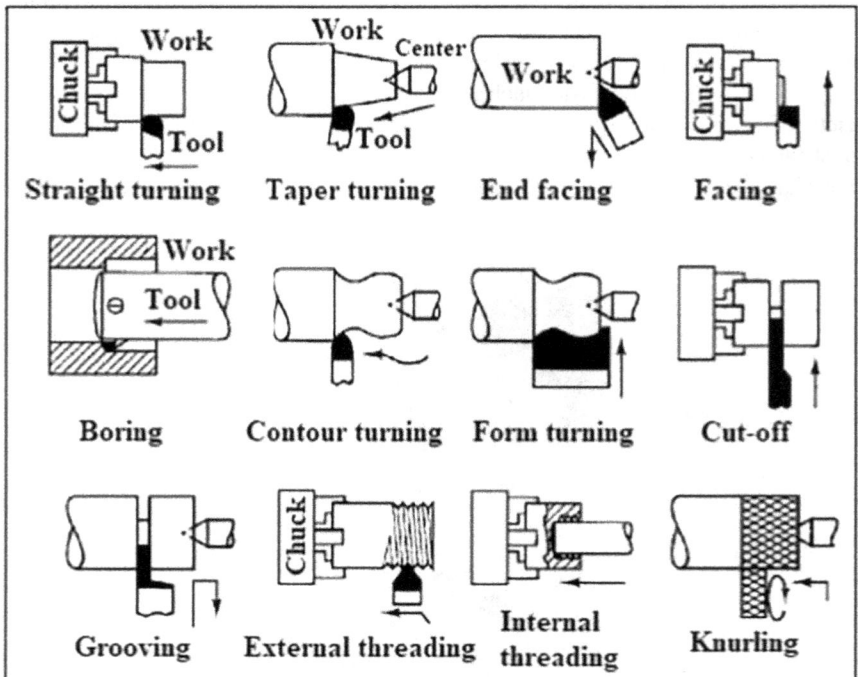

FIGURE 5.2   Commonly practiced turning-related operations in machining

In *straight turning*, the cutting tool moves longitudinally to produce straight cuts in the work; a rake angle of around 5° is recommended for the best cutting action in straight turning. In *taper turning*, there is a gradual reduction in diameter from one end of the cylindrical workpiece to the other end; this gradual reduction in diameter is accomplished by offsetting the tailstock of the lathe. In *facing* (or *face turning*), the tool is fed radially inward toward the center at the end of the work. In *boring*, a boring cutting tool is used to enlarge a hole usually made by a previous process. In *contour turning*, the tool follows a contour on the work thereby generating a contoured shape. In *form turning*, a form tool is fed perpendicular to the axis of rotation of the work to form the shape. In *cut-off turning*, the tool is fed perpendicular to the axis of rotation of the work at a location to cut-off end of the work. In *grooving*, the cutting tool moves longitudinally to produce a groove in the work. In *thread turning*, a pointed form tool is fed linearly across surface of work at a larger feed rate thus producing threads. In *knurling*, a pattern of straight, angled, or crossed lines is machined into the workpiece. Engineering analyses of some commonly practiced turning and related operations are presented in Section 5.5.

5.3 LATHE – *PARTS AND MECHANISM*

A *lathe* is equipped with a variety of inter-related parts and accessories, as illustrated in Figure 5.3. The size of a *lathe* is determined by the distance between the centers as well as by its *swing i.e.*, the diameter of work that can be put on a *lathe* (O'Brien and O'Brien, 2013). The *bed* supports all the other major components of the lathe. The *headstock* provides the driving mechanism and electrical connections in the lathe machine tool. The *headstock* holds the workpiece on its spindle through a work holding device (*e.g.*, a chuck); the spindle has a hole for securing work. The headstock transmits power from the spindle to the feed rod and lead screw. The *tool post* mounts tool holders in which cutting tools are clamped. The *compound rest* is

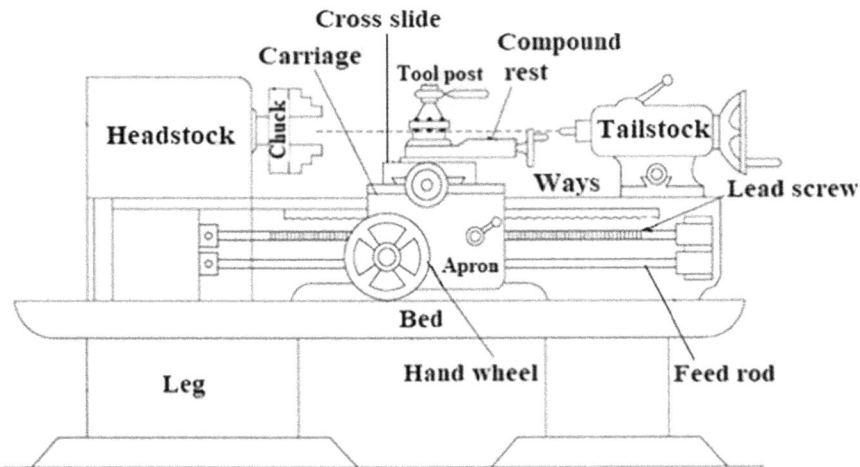

FIGURE 5.3   A *lathe* machine tool showing its major parts

mounted to the cross slide; it pivots around the tool post for positioning and adjust-ment of the cutting tool. The *hand wheel* is used to feed cutting tools into the work-piece. The *carriage* slides along the ways (see Figure 5.3). The *apron* is attached to the carriage and hangs over the front side of the lathe bed. The *feed rod* has a keyway with two reversing pinion gears to forward or reverse the carriage. The *lead screw* is used for cutting threads in the workpiece.

5.4 LATHE MACHINE TOOLS

Lathes are available in a variety of types and sizes. Based on their construction and use, lathes may be classified into the following types: (a) speed lathe, (b) engine lathe, (c) tool room lathe, (d) turret lathe, (e) automatic lathe, and (f) CNC lathes. These types of lathe, except CNC lathes, are briefly explained in the following sub-sections; the CNC lathes are discussed in Chapter 13.

5.4.1 SPEED LATHES

Speed lathes are the oldest and the simplest type of lathes. They are so named because of the very high speed at which the spindle rotates. They have most of the attach-ments which the other types of lathe carry but have no gear box, carriage, and the lead screw. *Speed lathes* have no provision for power feed; this is why the tool is fed and actuated by hand. They are usually employed for wood turning, polishing, cen-tring, metal spinning, and the like.

5.4.2 ENGINE LATHES

Engine lathes are the most commonly used type of lathes. The construction of an *engine lathe* is relatively more robust as compared to a *speed lathe*. The major parts of an engine lathe are illustrated in Figures 5.3 and 5.4. On an *engine lathe*, the tool is clamped onto a cross slide that is power driven on straight paths either perpendicu-lar or parallel to the workpiece axis. The *headstock* of an engine lathe incorporates suitable mechanism for providing multiple speeds to the lathe spindle (Walsh, 1994). The *headstock spindle* usually receives power from a motor; it carries a combination of gears. In modern engines lathes, there are provisions for longitudinal and cross feed movements by use of hand wheels as well as the *spindle speed selectors* situated in the headstock (see Figure 5.4). There are ON/OFF switches; it is important to make sure that the lathe is unplugged or isolated completely before making any adjustment during the machine operation. There is also an *emergency OFF* switch in case any need arises to immediately stop the machine.

5.4.3 TOOL ROOM LATHE

A *tool room lathe* is the lathe having a bed width greater than its swing, thereby mak-ing it more stable, rigid, and capable of higher accuracy as compared to an engine lathe. It is a high precision lathe having a gearbox in the headstock offering an extended range of thread pitches and feeds. It is used for making precision

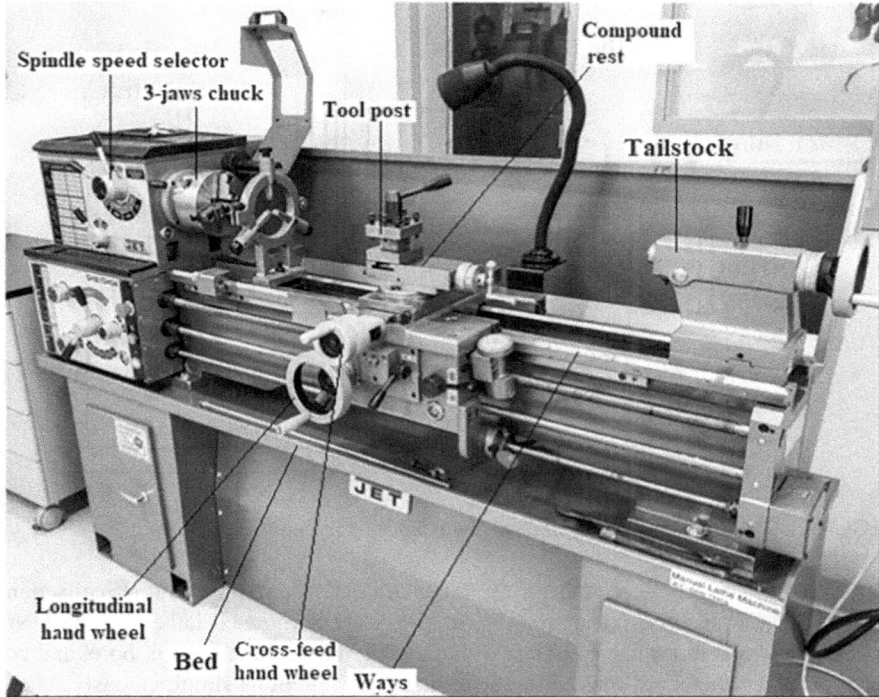

FIGURE 5.4 A modern engine lathe showing various parts of the machine tool

components with tight tolerances in the tool room; these components include tool holders, multiples of low-quantity hand-made precision screw threaded parts, and concentric parts with multiple steps, cuts, and holes.

5.4.4 TURRET LATHE

A *turret lathe* is the lathe that has multiple tools mounted on a turret; the latter is attached either to the tail stock or to the cross slide. Thus a *turret lathe* is so equipped as to enable a machinist to change cutting tools quickly. A standard lathe has a single tool post mounted on the cross slide. In order to change the tool in a standard lathe, one must unfasten the existing tool and replace it by another. A *turret lathe* has a mechanized turret to switch tools between operations within seconds. It is generally used for machining a work that will require different tools in a sequence, such as turning, facing, boring, thread turning, cut-off, and the like (Youssef and El-Hofy, 2008).

5.4.5 AUTOMATIC LATHE

Strictly speaking, the term "*automatic lathe*" used in machining refers to mechanically automated lathes usually controlled by cams. They differ from electronic

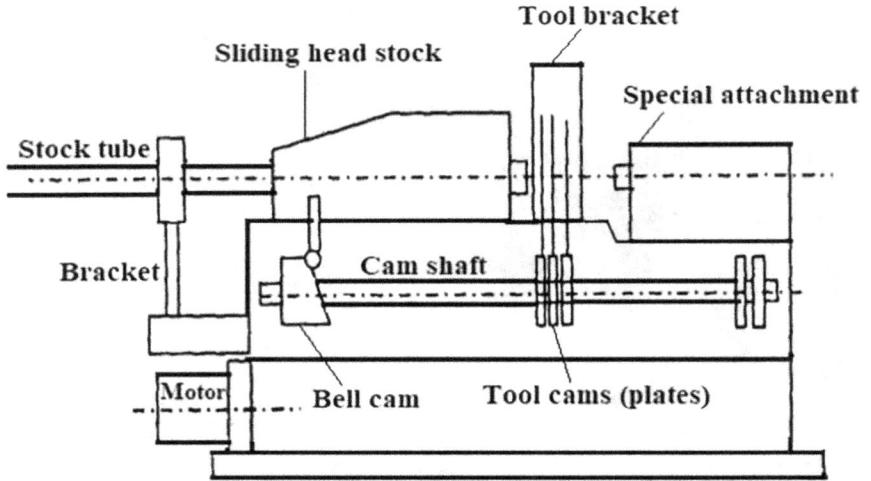

FIGURE 5.5   Automatic screw lathe machine (working principle)

automation via numerical control; the latter are called computer numerically controlled (CNC) machines. Thus *automatic lathes* are automated lathes of non-CNC types. These machines use spindles, which enable a machinist to turn, bore, and cut the workpiece thereby allowing to perform several functions simultaneously. There are two main types of automatic lathe: (a) automatic screw lathes and (b) automatic chucking lathe. *Automatic screw machines* are small- to medium-sized cam-operated automatic lathes; these machines work on workpieces/stock-tubes that are roughly up to 80 mm (3.1 in.) in diameter and around 300 mm (12 in.) in length (see Figure 5.5). *An automatic chucking lathe* is capable of handling larger work, which due to its size is more often chucking work and less often bar work.

5.5 WORK HOLDING TECHNIQUES IN LATHE PRACTICE

There are a number of work holding techniques in lathe machine tool practice; these techniques include (a) three-jaws chuck, (b) holding the work between centers, (c) collets, (d) face plate (see Figure 5.1), (e) mandrels, and the like. Since two of these techniques are generally practiced, they are discussed in the following paragraphs.

Three-Jaws Chuck. Workpiece is generally held in the headstock by use of a three-jaws chuck (see Figure 5.4). The three-jaws chuck work holding technique has the advantage that it enables the operator to set up the machine quickly, but by using this technique, the workpiece is not perfectly concentric. Additionally, sometimes there are issues related to jaw impressions made by chucks on a workpiece (Finkelstein et al., 2017). These technical problems can be overcome by using the technique of "turning between centers", as explained in the following paragraph.

Turning btween Centers. The *turning between centers* technique ensures that the work is perfectly concentric by "holding the work between centers". This technique involves the use of two centers: *dead center* and *live center*; the former is fitted to the

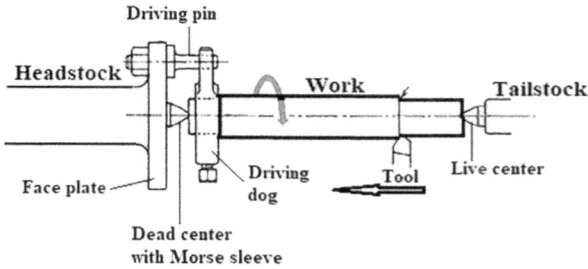

FIGURE 5.6 Turning between centers technique

headstock whereas the latter is fitted to the tailstock (see Figure 5.6). In order to fit the *dead center*, the chuck is removed and a Morse taper sleeve with *dead center* is attached to a face plate, which in turn is fitted to the headstock of the lathe. A *driving dog* is attached to the work which is driven by face plate driving pin. Finally, the *live center* is fitted to the tailstock (see Figure 5.6).

5.6 ENGINEERING ANALYSIS OF STRAIGHT TURNING

Straight turning operation can be analyzed by considering the three cutting conditions: cutting speed V, depth of cut d, and the feed f, as illustrated in Figure 5.7. The depths of cut, d, are generally in the range of 0.5–12 mm, whereas the feeds, f, lie in the range of 0.15–1 mm/rev (Kalpakjian and Schmid, 2008). Prior to the operation of a lathe, the machinist must select an appropriate cutting speed V based on the type of cutting tool and the work material to be machined by reference to the data in Table 5.1 (see Example 5.1).

Once the cutting speed for a particular set of work material and cutting tool has been selected, the spindle rpm (or work rotational speed) is calculated by using the following formula:

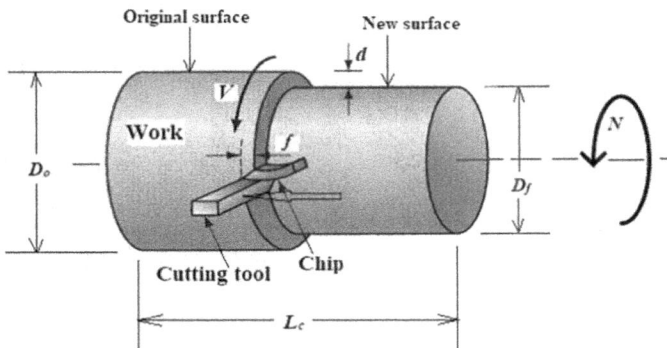

FIGURE 5.7 Straight turning (V = surface speed, f = feed, d = depth of cut, L_c = length of cut)

TABLE 5.1
Recommended Cutting Speeds for Turning Various Work Materials Using Tools

No.	Material	AISI/ASTM/SAE Designation	HSS Tool (ft/min)	Carbide Tool (ft/min)
1.	Aluminum (cast)	Sand and permanent mold casting alloys	600–750	2820
2.	Aluminum (wrought)	6061-T6, 5000, 6000, and 7000 series.	500–600	2820
3.	Brass	C35600, C37700, C36000, C33200, C34200, C35300, C48500, C34000	300–350	1170
4.	Bronze	C65500, C22600, C65100, C67500	200–250	715
5.	Cast iron	ASTM Class 20, 25, 30, 35, 40	145–215	410
6.	Cold work, air hardening tool steel	A2, A3, A4, A6, A7, A8, A9, A10	80–125	355–365
7.	Cold work, oil hardening tool steel	O1, O2, O6, O7	125	590
8.	Free machining plain carbon steels	1212, 1213, 1215	270–290	820–1045
9.	Free machining plain carbon steels	1108, 1109, 1115, 1117, 1118, 1120, 1126, 1211	215–235	950
10.	Free machining plain carbon steels	1132, 1137, 1139, 1140, 1144, 1146, 1151	70–215	670–800
11.	Free machining plain carbon steels (Leaded)	11L17, 11L18, 12L13, 12L14	200–260	800–820
12.	Plain carbon steels	1006, 1008, 1009, 1010, 1012, 1015, 1016, 1017, 1018, 1019, 1020, 1021, 1022, 1023, 1024, 1025, 1026, 1513, 1514	125–215	800–885
13.	Plain carbon steels	1027, 1030, 1033, 1035, 1036, 1037, 1038, 1039, 1040, 1041, 1042, 1043, 1045, 1046, 1048, 1049, 1050, 1052, 1524, 1526, 1527, 1541	55–180	670–970
14.	Free machining alloy steels (resulfurized)	4140, 4150	70–200	430–685
15.	Stainless steels (Austenitic)	201, 202, 301, 302, 304, 304L, 305, 308, 321, 347, 348	115–135	570
16.	Water hardening tool steel	W1, W2, W5	180	590

$$N = \frac{V}{\pi D_o} \tag{5.1}$$

where N is the spindle rpm or the work rotational speed, rev/min; V is the maximum cutting speed, mm/min; and D_o is the original work diameter, mm (see Example 5.2). Based on the computed spindle rpm, the machinist must set the *spindle speed selector* accordingly (see Figure 5.4).

It is a common practice to perform turning operation based on the average cutting speed. Thus Equation 5.1 can be modified as follows:

$$V_{av} = \pi D_{av} N \tag{5.2}$$

where V_{av} is the average cutting speed, D_{av} is the average work diameter, and N is the spindle rpm based on V_{av} and D_{av}. The average diameter (D_{av}) can be calculated as follows:

$$D_{av} = \frac{D_o + D_f}{2} \tag{5.3}$$

where D_o is the original work diameter, and D_f is the final work diameter (see Example 5.3).

Feed is related to the tool feed rate by:

$$f_r = f N \tag{5.4}$$

where f_r is the feed rate, mm/min, and f is the feed, mm/rev (see Example 5.4).

The depth of cut can be computed by using original and final diameters of the work, as follows:

$$d = \frac{D_o - D_f}{2} \tag{5.5}$$

where d is the depth of cut, D_o is the original work diameter, and D_f is the final work diameter.

The material removal rate can be computed as follows:

$$MRR = V d f \tag{5.6}$$

where MRR is the material removal rate, mm³/min (see Example 5.5).

The cutting time can be calculated by:

$$T_c = \frac{L_c}{f N} \tag{5.7}$$

where T_c is the cutting time, min, and L_c is the length of cut, mm (see Example 5.6).

In machine shop practice, there are often requirements in straight turning involving large depths of cuts; in such cases, it is necessary to perform turning operations in more than one cutting passes with small permissible depth of cut in a pass. In such a multiple cutting passes turning operation, the number of cutting passes is determined by:

$$\text{No. of cutting passes} = \frac{d}{d_{perm.}} \tag{5.8}$$

where $d_{perm.}$ is the permissible depth of cut in a pass. In such turning operations, it is useful to calculate the diameter after each cutting pass (see Example 5.10). In multiple cutting passes turning operations, the cutting time is the sum of individual cutting times of each pass taking into consideration the spindle rpm of the corresponding cutting pass (see Examples 5.11 and 5.12).

Once the MRR has been computed by Equation 5.6, the cutting power can be calculated by using Equation 3.9 (see Example 5.13). Then the cutting torque can by computed by:

$$Torq_c = \frac{P_c}{N_{(rad/s)}} \tag{5.9}$$

where $Torq_c$ is the cutting torque, N-m; P_c is the cutting power, W; and $N_{(rad/s)}$ is the work rotational speed, radians/s (see Example 5.14). Accordingly, the cutting force in turning can be calculated by using the following relationships (see Figure 3.1):

$$Torq_c = F_c \left(\frac{D_{av}}{2} \right) \tag{5.10}$$

By rearranging the terms in Equation 5.10, we obtain:

$$F_c = \frac{2 \left(Torq_c \right)}{D_{av}} \tag{5.11}$$

where F_c is the cutting force, N and D_{av} is the average diameter, m (see Example 5.15).

5.7 ENGINEERING ANALYSIS OF TAPER TURNING

Taper turning is a machining operation that involves the gradual reduction in diameter from one end of a cylindrical workpiece to the other end. *Taper turning* is performed on a lathe by offsetting its tailstock (see Figure 5.8a). In *taper turning*, a large diameter of taper "D" is gradually decreased to the small diameter of the taper "d" at the horizontal length "l" of tapered part of the workpiece (Figure 5.8b). In Figure 5.8b, the length $PQ = (D - d)/2$.

The *taper angle* (also called *cone angle*) is related to the large and small diameters by:

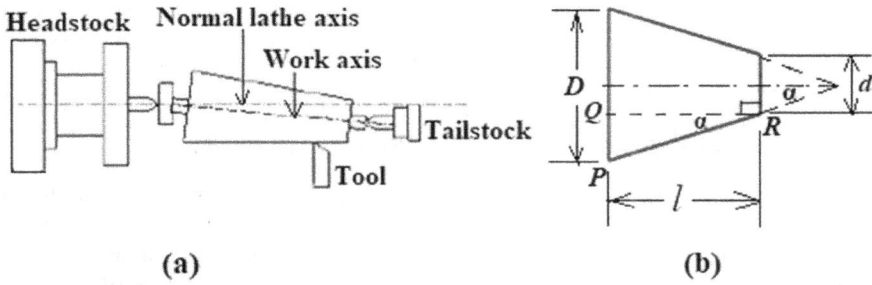

FIGURE 5.8 Taper turning operation: (a) offsetting of the tail-stock, (b) engineering analysis

$$\tan \alpha = \frac{\overline{PQ}}{l} = \frac{D-d}{2l} \qquad (5.12)$$

where α is the half of the taper angle, deg.; D is the large diameter, d is the small diameter, and l is the horizontal length of the tapered part of the workpiece (see Example 5.18).

Taper ratio is the ratio of 1 mm to the unit length of the taper. In Figure 5.9, "k" is the unit length of the taper so the taper ratio = $1/k$. The large diameter D is related to d, l, and k by:

$$D = d + \frac{l}{k} \qquad (5.13)$$

The significance of Equation 5.10 is illustrated in Example 5.19.

The amount of *offset* for the tapered workpiece turned by offsetting the tailstock (see Figure 5.8a and Figure 5.9) can be computed by:

$$Offset = \frac{L(D-d)}{2l} \qquad (5.14)$$

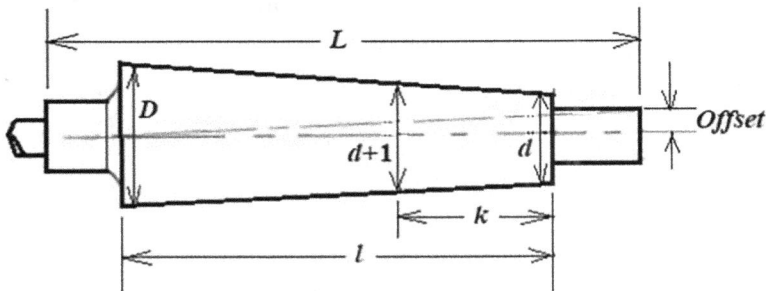

FIGURE 5.9 The tapered length of work showing unit length of taper and the offset

where D and d are the large and small diameters, respectively; l is the horizontal length of the tapered part of the work; and L is the overall length of the workpiece (see Example 5.20).

5.8 ENGINEERING ANALYSIS OF THREAD TURNING

Thread turning (or *threading*) involves the use of a V-shape-tipped-form tool that is fed linearly at a large feed rate across the surface of a rotating workpiece to produce *threads* (see Figure 5.10). In *threading*, the tool feed rate must correspond precisely to the pitch of the thread. Each thread making requires one revolution of the work. Thus, for example, if there are 4 threads per cm, the tool has to travel at a feed rate of 4 revolution per cm *i.e.*, the feed must be 0.25 or 2.5 mm/rev. By comparison with conventional turning application, the feed in thread turning may be ten times greater than normal turning. *Thread turning* is more demanding than normal turning operations owing to the wide applications of threaded screws, bolts, and other threaded parts.

It is evident in Figure 5.10 that the pitch of the threaded section of the work can be calculated by:

$$\text{Pitch} = \frac{\text{Length of cut}}{\text{Number of threads}} \tag{5.15}$$

Since each thread making requires one complete revolution of the work, the number of revolutions required in thread turning can be computed by:

$$\text{Number of revolutions required} = \frac{\text{Length of cut}}{\text{Pitch}} \tag{5.16}$$

The significance of Equations 5.15 and 5.16 is illustrated in Examples 5.21 and 5.22.

It is explained in Section 5.4 that thread turning requires a working mechanism with the *lead screw* of the lathe (see Figure 5.3). Since the headstock spindle is linked with a driver gear, the *lead screw* is attached to a driven gear through an idle gear. In thread turning, the lathe machine tool is so designed that the gear ratio can be computed by using the following formula:

$$\text{GR} = \frac{\text{Number of teeth on the driven gear}}{\text{Number of teeth on the driver gear}}$$

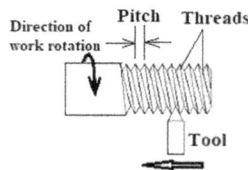

FIGURE 5.10   Thread turning operation showing directions of tool and work motions

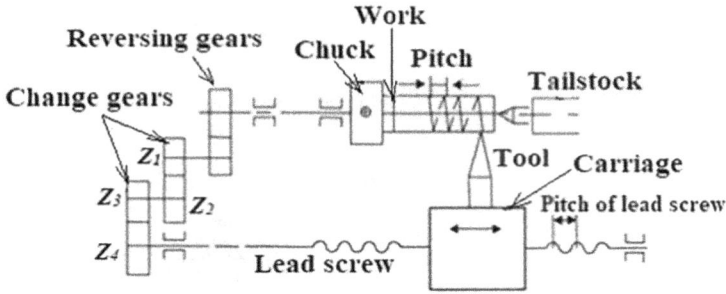

FIGURE 5.11 Thread turning mechanism with a change gear quadrant (Z_1 and Z_3 indicate teeth on driver gears, whereas Z_2 and Z_4 indicate teeth on driven gears)

or

$$GR = \frac{\text{Number of teeth on the leadscrew gear}}{\text{Number of teeth on the work gear}} = \frac{\text{TPI on the leadscrew gear}}{\text{TPI on the work gear}} \quad (5.17)$$

where GR is the gear ratio, and TPI is the teeth per inch (see Example 5.23).

In lathe thread-cutting practice, there is often a need of using "change gears" for threading. The *change gears* refer to a series of gears that drive the lead screw. They are so named because we change them (based on the number of teeth on them) for turning to obtain different thread pitches. Figure 5.11 illustrates the thread turning mechanism with change gears.

By reference to Equation 5.17, the change gear ratio (GR_{cg}) can be computed by:

$$GR_{cg} = \frac{\text{Pitch of the work}}{\text{Pitch of the lead screw}} = \frac{Z_1}{Z_2} \times \frac{Z_3}{Z_4} \quad (5.18)$$

where Z_1, Z_2, Z_3, and Z_4 are the number of teeth on the change gears (see Figure 5.11). In case of simple gearing, just two gears (Z_1 and Z_2) are used (see Examples 5.23 and 5.24), whereas in compound gearing, all four change gears are used (see Figure 5.11, Example 5.25).

5.9 CALCULATIONS – WORKED EXAMPLES ON TURNING OPERATIONS AND MACHINES

EXAMPLE 5.1: SELECTING THE CUTTING SPEED FOR STRAIGHT TURNING OPERATION

A cylindrical AISI-1115 free machining plain carbon steel bar is to be turned on a lathe. Select the maximum cutting speed (in mm/min) for (a) HSS tool and (b) carbide tool.

Solution

a. By reference to Table 5.1, cutting speed for HSS tool = V = 235 ft/min

$$V = 235\frac{\text{feet}}{\text{min}} = \frac{235 \times 12\,\text{in.}}{\text{min}} = \frac{235 \times 12 \times 25.4\,\text{mm}}{\text{min}}$$
$$= \frac{235 \times 304.8\,\text{mm}}{\text{min}} = 71,628\,\text{mm/min}$$

b. By reference to Table 5.1, cutting speed for carbide tool = V = 950 ft/min

$$V = 950\frac{\text{feet}}{\text{min}} = \frac{950 \times 304.8\,\text{mm}}{\text{min}} = 289,560\,\text{mm/min}$$

EXAMPLE 5.2: COMPUTING THE SPINDLE RPM AND SETTING THE SPINDLE SPEED SELECTOR

An AISI-1115 free machining plain carbon steel bar with an original diameter of 14 mm is to be turned by using a HSS tool. Calculate the spindle rpm, and hence set the *spindle speed selector* out of the following three options in the lathe: 1000, 1500, and 2000 rpm.

Solution
By reference to Example 5.1, V = 71,628 mm/min, D_o = 14 mm, N = ?
By using Equation 5.1,

$$N = \frac{V}{\pi D_o} = \frac{71,628}{14\pi} = 1628.35\,\text{rev/min}$$

The spindle rpm = 1628
The spindle speed selector must be set on "1500 rpm".

EXAMPLE 5.3: COMPUTING SPINDLE RPM BASED ON AVERAGE CUTTING SPEED AND AVG. DIAMETER

A *C65100 bronze* workpiece is to be straight turned by reducing its diameter from 18 to 15 mm by using an HSS tool. Calculate the work rotational speed (or spindle rpm) based on the average cutting speed and average diameter.

Solution
By reference to Table 5.1, $V_{av} = \frac{200 + 250}{2} = 225$ ft/min = 225 × 304.8 = 68,580 mm/min
By using Equation 5.3,

$$D_{av} = \frac{D_o + D_f}{2} = \frac{18 + 15}{2} = 16.5\,\text{mm}$$

By using the modified form of Equation 5.2,

$$N = \frac{V_{av}}{\pi D_{av}} = \frac{68,580}{16.5\pi} = 1322.8\,\text{rev/min}$$

The work rotational speed = 1323 rev/min.

EXAMPLE 5.4: COMPUTING THE FEED IN STRAIGHT TURNING FOR ALUMINUM ALLOY

A 7075 aluminum alloy 80-cm-long bar with a diameter of 18 mm is to be machined by straight turning to a diameter of 13 mm. The cutting tool is made of cemented carbide, and the tool feed rate is 1 m/min. Calculate the feed.

Solution
D_o = 18 mm, f_r = 1 m/min = 1000 mm/min, f = ?
  By reference to Table 5.1, V = 2820 ft/min = 2820 × 304.8 = 859,536 mm/min.
  By using Equation 5.1,

$$N = \frac{V}{\pi D_O} = \frac{859,536}{18\,\pi} = 15,198\,\text{rev/min}$$

By using Equation 5.4,

$$f = \frac{f_r}{N} = \frac{1000}{15,198} = 0.065\,\text{mm/rev}$$

The feed = f = 0.065 mm/rev.

EXAMPLE 5.5: COMPUTING THE DEPTH OF CUT AND MATERIAL REMOVAL RATE

By using the data in Example 5.4, calculate the material removal rate.

Solution
D_o = 18 mm, D_f = 13 mm, V = 859,536 mm/min, f = 0.065 mm/rev., d = ?, MRR = ?
  By using Equation 5.5,

$$d = \frac{D_o - D_f}{2} = \frac{18-13}{2} = 2.5\,\text{mm}$$

By using Equation 5.6,

$$\text{MRR} = V\,d\,f = 859,536 \times 2.5 \times 0.065 = 139,675\,\text{mm}^3/\text{min}$$

MRR = 139,675 mm³/min

EXAMPLE 5.6: CALCULATING THE CUTTING TIME IN STRAIGHT TURNING

By using the data in Example 5.4, calculate the cutting time for the turning operation.

Solution
L_c = 80 cm = 800 mm, f = 0.065 mm/rev, N = 15,198 rev/min, T_c = ?

By using Equation 5.7,

$$T_c = \frac{L_c}{f\,N} = \frac{800}{0.065 \times 15,198} = 0.81\,\text{min}$$

The cutting time = 0.81 min or 49 s.

EXAMPLE 5.7: COMPUTING/SELECTING THE CUTTING SPEED WHEN WORK MATERIAL IS UNKNOWN

A 35-cm-long cylindrical workpiece is required to be machined by straight turning in 3 min. The original diameter of the work is 70 mm. What cutting speed do you recommend for turning the work by using a feed of 0.28 mm/rev, and depth of cut of 3 mm?

Solution
L_c = 35 cm = 350 mm, T_c = 3 min, D_o = 70 mm, f = 0.28 mm/rev, d = 3 mm, V = ?
By using modified form of Equation 5.7,

$$N = \frac{L_c}{T_c f} = \frac{350}{3 \times 0.28} = 416.67\,\text{rev/min}$$

By using the modified form of Equation 5.1,

$$V = \pi\,D_o\,N = 3.142 \times 70 \times 416.67 = 91,641\,\text{mm/min} = 91.641\,\text{m/min}$$

The cutting speed = 91.641 m/min.

EXAMPLE 5.8: IDENTIFYING THE MACHINING DATA FROM THE SKETCH OF TURNING OPERATION

By reference to Figure E-5.8, identify the following machining data: (a) original work diameter, (b) final work diameter, (c) cutting speed, (d) tool feed rate, and (e) length of cut.

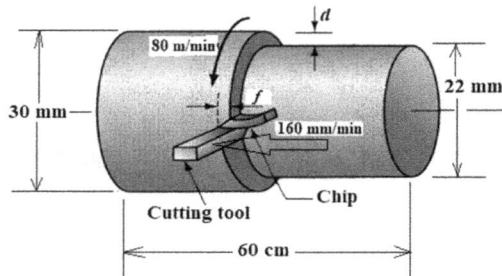

FIGURE E-5.8   Schematic showing straight turning operation's data

Solution

By reference to Figure E-5.8,
- a. Original work diameter $= D_o = 30$ mm
- b. final work diameter $= D_f = 22$ mm
- c. Tool feed rate $= f_r = 160$ mm/min
- d. Cutting speed $= V = 80$ m/min $= 80,000$ mm/min
- e. Length of cut $= L_c = 60$ cm $= 600$ mm

EXAMPLE 5.9: CALCULATING THE FEED, MRR, AND CUTTING TIME

By using the data in Example 5.8, calculate the feed, MRR, and cutting time.

Solution

$D_o = 30$ mm, $D_f = 22$ mm, $f_r = 160$ mm/min, $V = 80,000$ mm/min, $L_c = 600$ mm, $f = ?$, $MRR = ?$, $T_c = ?$

By using Equation 5.1,

$$N = \frac{V}{\pi D_o} = \frac{80,000}{30\pi} = 848.72 \text{ rev/min}$$

By using Equation 5.4,

$$f = \frac{f_r}{N} = \frac{160}{848.72} = 0.188 \text{ mm/rev}$$

By using Equation 5.5,

$$d = \frac{D_o - D_f}{2} = \frac{30 - 22}{2} = 4 \text{ mm}$$

By using Equation 5.6,

$$MRR = V\, d\, f = 80,000 \times 4 \times 0.188 = 60,160 \text{ mm}^3/\text{min}$$

By using Equation 5.7,

$$T_c = \frac{L_c}{f\, N} = \frac{600}{0.188 \times 848.72} = 3.76 \text{ min}$$

Feed $= 0.188$ mm/rev, MRR $= 60160$ mm³/min, Cutting time $= 3.76$ min.

EXAMPLE 5.10: DETERMINING NUMBER OF CUTTING PASSES AND THE DIAMETER AFTER EACH PASS

A C48500 brass bar is straight turned by reducing its diameter from 60 to 50 mm by using an HSS tool. The permissible depth of cut is 2.5 mm. The cutting length of the bar is 50 cm. The feed is 0.6 mm/rev. Determine the (a) number of cutting passes and (b) the diameter after each cutting pass by drawing a sketch of the operation showing the diameter(s) after each pass.

Solution

$D_o = 60$ mm, $D_f = 50$ mm, $L_c = 50$ cm $= 500$ mm, $d_{perm.} = 2.5$ mm, $f = 0.6$ mm/rev.

   a. By using Equation 5.5,

$$d = \frac{D_o - D_f}{2} = \frac{60 - 50}{2} = 5\,\text{mm}$$

By using Equation 5.8,

$$\text{No. of cutting passes} = \frac{d}{d_{perm.}} = \frac{5}{2.5} = 2$$

   b. For the 1st Cutting Pass: $d_{perm.} = \frac{D_o - D_{f(1)}}{2}$ or $2.5 = \frac{60 - D_{f(1)}}{2}$

$$60 - D_{f(1)} = 5$$

$$D_{f(1)} = 55\,\text{mm}$$

The diameter after the first cutting pass $= D_{f(1)} = 55$ mm

   For the 2nd Cutting Pass:

   The initial diameter for the 2nd pass = the diameter after the 1st pass (see Figure E-5.10)

$$D_{o(2)} = D_{f(1)}$$

$$d_{perm.} = \frac{D_{o(2)} - D_f}{2} \quad \text{or} \quad 2.5 = \frac{55 - D_f}{2}$$

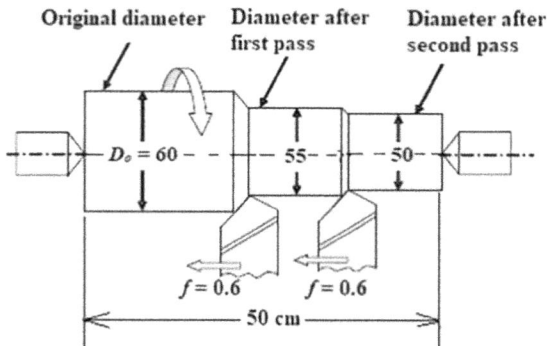

FIGURE E-5.10   Sketch of turning operation showing diameter(s) after each cutting pass

$$55 - D_f = 5$$

$$D_f = 50\,\text{mm}$$

The sketch of the operation showing the diameter(s) after each pass is shown in Figure E-5.10.

EXAMPLE 5.11: COMPUTING SPINDLE RPM FOR MULTIPLE CUTTING PASSES TURNING OPERATION

By using the data in Example 5.10, calculate the work rotational speed (spindle rpm) for each cutting pass of the turning operation.

Solution
By reference to Table 5.1, the maximum cutting speed for turning C48500 brass bar by using an HSS tool is 350 ft/min *i.e.*, $V = 350$ ft/min $= 350 \times 304.8 = 106{,}680$ mm/min.
  For the 1st Cutting Pass: $D_{o(1)} = 60$ mm, $V = 106{,}680$ mm/min
  By using Equation 5.1 for the 1st cutting pass,

$$N_1 = \frac{V}{\pi\,D_{o(1)}} = \frac{106{,}680}{60\,\pi} = 565.88\,\text{rev/min}$$

The work rotational speed for the 1st cutting pass = 565.88 rev/min
  For the 2nd Cutting Pass: $D_{o(2)} = 55$ mm, $V = 106{,}680$ mm/min

$$N_2 = \frac{V}{\pi\,D_{o(2)}} = \frac{106{,}680}{55\,\pi} = 617.32\,\text{rev/min}$$

The work rotational speed for the 2nd cutting pass = 617.32 rev/min.

EXAMPLE 5.12: COMPUTING CUTTING TIME FOR MULTIPLE CUTTING PASSES TURNING OPERATION

By using the data in Example 5.11, calculate the total cutting time for the multiple cutting passes turning operation.

Solution
$L_c = 50$ cm $= 500$ mm, $f = 0.6$ mm/rev, $N_{(1)} = 565.88$ rev/min, $N_{(2)} = 617.32$ rev/min.
  By using Equation 5.7 for the 1st cutting pass,

$$T_{c(1)} = \frac{L_c}{f\,N_{(1)}} = \frac{500}{0.6 \times 565.88} = 1.47\,\text{min}$$

By using Equation 5.7 for the 2nd cutting pass,

$$T_{c(2)} = \frac{L_c}{f\,N_{(2)}} = \frac{500}{0.6 \times 617.32} = 1.35\,\text{min}$$

Total cutting time = $T_c = T_{c(1)} + T_{c(2)} = 1.47 + 1.35 = 2.82$ min.

EXAMPLE 5.13: CALCULATING THE CUTTING POWER IN TURNING WHEN THE MRR IS NOT GIVEN

A 7075 aluminum alloy bar with a diameter of 18 mm is being machined by straight turning to a diameter of 13 mm. The cutting tool is made of cemented carbide. The feed is 0.12 mm/rev. Calculate the cutting power.

Solution
D_o = 18 mm, D_f = 13 mm, f = 0.12 mm/rev, P_c = ?
    By reference to Table 5.1, V = 2820 ft/min = 2820 × 304.8 = 859,536 mm/min
    By using Equation 5.5,

$$d = \frac{D_o - D_f}{2} = \frac{18-13}{2} = 2.5\,\text{mm}$$

By using Equation 5.6,

$$MRR = V\,d\,f = 859,536 \times 2.5 \times 0.12 = 257,860.8\,\text{mm}^3/\text{min} = 4297.68\,\text{mm}^3/\text{s}$$

By reference to Table 3.1, for aluminum alloys, P_u = 0.8 W · s/mm³.
    By using Equation 3.9,

$$P_u = \frac{P_c}{MRR}$$

$$P_c = (P_u)(MRR) = (0.8\,\text{W}\cdot\text{s/mm}^3) \times (4297.68\,\text{mm}^3/\text{s}) = 3438.14\,\text{W}$$

The cutting power in the turning operation = P_c = 3438.14 W.

EXAMPLE 5.14: CALCULATING THE CUTTING TORQUE IN TURNING OPERATION

By using the data in Example 5.14, calculate the cutting torque in the turning operation.

Solution
D_o = 18 mm, D_f = 13 mm, V = 859,536 mm/min, P_c = 3438.14 W, $Torq_c$ = ?
    By using Equation 5.1,

$$N = \frac{V}{\pi\,D_o} = \frac{859,536}{18\pi} = 15,198\,\text{rev/min}$$

$$N_{(rad/s)} = 15,198 \times 2\pi = 95,504.2 \, \text{radians/min} = 1591.73 \, \text{rad/s}$$

By using Equation 5.9,

$$Torq_c \frac{P_c}{N_{(rad/s)}} = \frac{3438.14}{1591.73} = 2.16 \, \text{N} - \text{m}$$

The cutting torque in the turning operation = $Torq_c$ = 2.16 N-m.

EXAMPLE 5.15: DETERMINING THE CUTTING FORCE WITHOUT USING A DYNAMOMETER

By using the data in Examples 5.13 and 5.14, calculate the cutting force in the turning operation.

D_o = 18 mm, D_f = 13 mm, $Torq_c$ = 2.16 N-m, F_c = ?

By using Equation 5.3,

$$D_{av} = \frac{D_o + D_f}{2} = \frac{18 + 13}{2} = 15.5 \, \text{mm} = 0.015 \, \text{m}$$

By using Equation 5.11,

$$F_c = \frac{2 \left(Torq_c \right)}{D_{av}} = \frac{2 \times 2.16}{0.015} = 288 \, \text{N}$$

The cutting force is the turning operation = 288 N.

EXAMPLE 5.16: CALCULATING THE FEED WHEN THE TOOL FEED RATE IS UNKNOWN

A 1-m-long W1 water hardened tool steel bar with 7 cm diameter is chucked in an engine lathe and supported at the opposite end using a live center. A length of 88 cm is to be turned to a diameter of 4 cm by using a carbide tool. The MRR should be 100 cm³/min. Calculate the required (a) depth of cut and (b) feed.

Solution

By reference to Table 5.1, V = 590 ft/min = 590 × 12 × 2.54 = 17,983 cm/min.

D_o = 7 cm, D_f = 4 cm, MRR = 100 cm³/min, d = ?, f = ?

a. By using Equation 5.5,

$$d = \frac{D_o - D_f}{2} = \frac{7 - 4}{2} = 1.5 \, \text{cm}$$

b. By using modified form of Equation 5.6,

$$f = \frac{MRR}{Vd} = \frac{100}{17,983 \times 1.5} = 0.0037 \, \text{cm/rev} = 0.037 \, \text{mm/rev}$$

The depth of cut is 15 mm and the feed is 0.037 mm/rev.

EXAMPLE 5.17: CALCULATING THE CUTTING TIME WHEN THE SPINDLE RPM IS NOT GIVEN

By using the data in Example 5.16, calculate the cutting time for the single-pass turning operation.

Solution
$L_c = 88$ cm, $D_o = 7$ cm, $f = 0.0037$ cm/rev, $V = 17,983$ cm/min, $T_c = ?$
    By using Equation 5.1,

$$N = \frac{V}{\pi D_o} = \frac{17,983}{7\pi} = 817.6 \, \text{rev/min}$$

By using Equation 5.7,

$$T_c = \frac{L_c}{f N} = \frac{88}{0.0037 \times 817.6} = 29 \, \text{min}$$

The cutting time = 29 min.

EXAMPLE 5.18: CALCULATING THE TAPER ANGLE FOR TAPER TURNING OF A WORKPIECE

A large diameter (= 20 cm) of taper is gradually decreased to the small diameter (= 12 cm) of the taper across the horizontal length (= 75 cm) of the tapered part of the workpiece. Calculate the taper angle.

Solution
$D = 20$ cm, $d = 12$ cm, $l = 75$ cm, taper angle $= 2\alpha = ?$
    By using Equation 5.12,

$$\tan \alpha = \frac{D - d}{2l} = \frac{20 - 12}{2 \times 75} = 0.053$$

$$\alpha = 3°$$

The taper angle $= 2\alpha = 6°$.

EXAMPLE 5.19: CALCULATING THE TAPER RATIO FOR A TAPERED WORKPIECE

The large diameter of the tapered section of a 280-mm-long workpiece is 19.6 mm. The small diameter of the tapered section is 17 mm. Calculate the taper ratio for 65 mm length on the work.

Solution
l = 65 mm, d = 17 mm, D = 19.6 mm, 1:k = ?
  By using Equation 5.13,

$$19.6 = 17 + \frac{65}{k}$$

$$\frac{65}{k} = 19.6 - 17$$

$$k = \frac{65}{2.6} = 25$$

Taper ratio = 1:k = 1:25.

EXAMPLE 5.20: CALCULATING THE TAILSTOCK OFFSET IN TAPER TURNING

By using the data in Example 5.19, calculate the tailstock offset required to turn the workpiece.

Solution
L = 280 mm, l = 65 mm, d = 17 mm, D = 19.6 mm, Offset?
  By using Equation 5.14,

$$Offset = \frac{L(D-d)}{2l} = \frac{280(19.6-17)}{2 \times 65} = 5.6\,mm$$

The tailstock offset = 5.6 mm.

EXAMPLE 5.21: CALCULATING THE PITCH OF A THREADED BOLT

A threaded bolt has 40 threads in 3 cm length of cut. Calculate the pitch of the bolt. How many revolutions of the bolt were performed in the thread turning operation?

Solution
Length of cut = 3 cm, No. of threads = 40, Pitch = ?
  By using Equation 5.15,

$$Pitch = \frac{Length\ of\ cut}{Number\ of\ threads} = \frac{3\,cm}{40} = 0.075\,cm = 0.75\,mm$$

Number of revolutions of the bolt = Number of threads on the bolt = 40.

EXAMPLE 5.22: COMPUTING THE NUMBER OF REVOLUTIONS OF WORK AND LEAD SCREW IN THREADING

An engine lathe has a lead screw with a pitch of 6 mm. It is required to cut threads with 1.1-mm pitch across a length of 60 mm in a bolt. Calculate the number of revolution required for the (a) bolt and (b) lead screw.

Solution
By using Equation 5.16,

$$\text{Number of revolutions of the bolt} = \frac{\text{Length of cut}}{\text{Pitch of the bolt threads}}$$
$$= \frac{60\,\text{mm}}{1.1\,\text{mm}} = 54\,\tfrac{1}{2}$$

$$\text{Number of revolutions of the lead screw} = \frac{\text{Length of cut}}{\text{Pitch of the lead screw threads}}$$
$$= \frac{60\,\text{mm}}{6\,\text{mm}} = 10$$

EXAMPLE 5.23: CALCULATING THE GEAR RATIO IN THREAD TURNING

Twelve threads to an inch are required to be cut in a bolt. The lead screw has 18 threads per inch. Calculate the gear ratio.

Solution
TPI in lead screw = 18, TPI in work = 12, Gear ratio = ?
By using Equation 5.17,

$$\text{Gear ratio} = \frac{\text{TPI on the leadscrew gear}}{\text{TPI on the work gear}} = \frac{18}{12} = 1.5$$

EXAMPLE 5.24: CALCULATING THE CHANGE GEAR RATIO FOR SIMPLE GEARING IN THREADING

A single start thread of 2-mm pitch is required to be cut in a work on a lathe having a lead screw pitch of 6 mm. Calculate the change gear ratio.
The available change gears are 20, 25, 30, 75, and 85.

Solution
Pitch of the work = 2 mm, Pitch of the lead screw = 6 mm, GR_{cg} = ?
By using the simplified form of Equation 5.18,

$$GR_{cg} = \frac{\text{Pitch of the work}}{\text{Pitch of the lead screw}} = \frac{Z_1}{Z_2} = \frac{2\,\text{mm}}{6\,\text{mm}} = \frac{1}{3} = \frac{1 \times 25}{3 \times 25} = \frac{25}{75}$$

The gear ratio is as follows:
The number of teeth on the driver gear $Z_1 = 25$.
The number of teeth on the driven gear $Z_2 = 75$.

EXAMPLE 5.25: CALCULATING THE CHANGE GEAR RATIO FOR COMPOUND GEARING THREADING

Calculate the change gear ratio for cutting a single-start thread of 0.25-mm pitch on a lathe having a lead screw pitch of 6 mm.
The available change gears are 20, 25, 30, 40, 50, 100, 120, 150.

Solution
Pitch of the work = 0.25 mm, Pitch of the lead screw = 6 mm, $GR_{cg} = ?$
By using Equation 5.18,

$$GR_{cg} = \frac{\text{Pitch of the work}}{\text{Pitch of the lead screw}} = \frac{Z_1}{Z_2} \times \frac{Z_3}{Z_4} = \frac{0.25\,\text{mm}}{6\,\text{mm}} = \frac{0.25 \times 4}{6 \times 4} = \frac{1}{24}$$

or

$$\frac{Z_1}{Z_2} \times \frac{Z_3}{Z_4} = \frac{1}{24} = \frac{1}{6} \times \frac{1}{4} = \frac{1 \times 20}{6 \times 20} \times \frac{1 \times 25}{4 \times 25} = \frac{20}{120} \times \frac{25}{100}$$

The driver gears are $Z_1 = 20$ teeth, $Z_3 = 25$ teeth.
The driven gears are $Z_2 = 120$ teeth, $Z_4 = 100$ teeth.

QUESTIONS AND PROBLEMS

5.1   Encircle the best answers for the following statements.
   a. Which type of lathe has a bed width greater than its swing?
       (i) Engine lathe, (ii) tool room lathe, (iii) turret lathe, (iv) automatic lathe
   b. Which type of lathe is the best for quick change of cutting tools during operation?
       (i) Engine lathe, (ii) tool room lathe, (iii) turret lathe, (iv) automatic lathe
   c. Which type of lathe is most commonly used in machine shop practice?
       (i) Engine lathe, (ii) tool room lathe, (iii) turret lathe, (iv) automatic lathe
   d. Which type of lathe involves the use of cam shaft in its working mechanism?
       (i) Engine lathe, (ii) tool room lathe, (iii) turret lathe, (iv) automatic lathe
   e. In which type of turning operation is lead screw of a lathe used?
       (i) straight turning, (ii) taper turning, (iii) thread turning, (iv) contour turning.

f. Which type of turning operation involves offset of tool post?
(i) straight turning, (ii) taper turning, (iii) thread turning, (iv) contour turning.
g. In which type of turning, a pattern of crossed lines is machined into the workpiece?
(i) knurling, (ii) grooving, (iii) facing, (iv) cut-off, (v) threading
h. Which type of turning involves the use of a form tool?
(i) knurling, (ii) grooving, (iii) facing, (iv) cut-off, (v) threading
i. Which part of a lathe slides along the ways during turning operation?
(i) carriage, (ii) tailstock, (iii) headstock, (iv) chuck, (v) tool

5.2 Draw a sketch of an engine lathe showing its major parts.

5.3 Select the turning operation to produce the following parts, and draw a sketch of the corresponding turning operation: (a) Roll for rolling bars, (b) screw, (c) shaft with different diameters at the ends, (d) body of micrometer screw gage.

5.4 Compare the capabilities of the following machine tools: (a) engine lathe, (b) tool room lathe.

5.5 Draw a sketch of threading operation showing pitch, length of cut, and other motion details.

5.6 Differentiate between the following machine tools: (a) automatic screw lathe, (b) automatic chucking lathe.

P5.7 A cast iron workpiece is to be straight turned by reducing its diameter from 22 to 17 mm by using an HSS tool. Calculate the work rotational speed (or spindle rpm) based on the average cutting speed and average diameter.

P5.8 A straight turning operation is performed using the following machining data:
Original work diameter = 40 mm, Machined work diameter = 30 mm,
Feed rate = 190 mm/min, Cutting speed = 90 m/min, Length of cut = 70 cm.
a. Draw a sketch of the turning operation showing the machining data.
b. Calculate the (i) feed, (ii) materials removal rate, and (iii) cutting time.

P5.9 In a straight turning operation for brass bar with 30 cm length, the diameter is to be reduced from 100 to 94 mm. The feed is 0.2 mm/rev, and the cutting speed is 45 m/min. The permissible depth of cut is 1.5 mm. Calculate the (a) number of cutting passes, (b) the spindle rpm for each pass, and (c) cutting time for the turning operation. Draw a sketch of the operation showing the diameter(s) after each pass.

P5.10 A 304 austenitic stainless steel bar with a diameter of 20 mm was machined by turning to a diameter of 14 mm. The cutting tool is made of cemented carbide. The feed was 0.14 mm/rev. Calculate the (a) cutting power, (b) cutting torque, and (c) cutting force.

P5.11 A large diameter (= 30 mm) of taper is gradually decreased to the small diameter (= 13 mm) of the taper across the horizontal length (= 20 cm) of the tapered part of the work. Calculate the taper angle.

P5.12 Calculate the change gear ratio for cutting a single-start thread of 0.5-mm pitch on a lathe having a lead screw pitch of 18 mm. The available change gears are 17, 20, 80, 100, 140, 153, 160

P5.13 The large diameter of the tapered section of a 260-mm-long workpiece is 22 mm. The small diameter of the tapered section is 18 mm. Calculate the (a) taper ratio for 60 mm length on the work and (b) tailstock offset.

P5.14 An engine lathe has a lead screw with a pitch of 5 mm. It is required to cut threads with 1.33-mm pitch across a length of 80 mm in a bolt. Calculate the number of revolution required for the (a) bolt and (b) lead screw.

REFERENCES

Finkelstein, N., Aronson, A., Tsach, T., (2017), Toolmarks made by lathe chuck jaws, *Forensic Science International*, 275, 124–127.
Kalpakjian, S., Schmid, S. (2008), *Manufacturing Processes for Engineering Materials* (5th Edition), Pearson Education plc, London, UK.
O'Brien, J.J., O'Brien, M.W. (2013), *How to Run a Lathe: The Care and Operation of a Screw Cutting Lathe*, Martino Fine Books, Eastford, CT.
Walsh, R.A. (1994), *Machining and Metalworking Handbook*, McGraw Hill, New York.
Yates, R.F. (1922), *Lathe Work for Beginners – A Practical Treatise*, Norman W. Henley Publishing Company, New York.
Youssef, H.A., El-Hofy, H. (2008), *Machining Technology: Machine Tools and Operations*, CRC Press, Boca Raton, FL.

6 Drilling Operations and Machines

6.1 DRILLING AND ITS INDUSTRIAL APPLICATIONS

Drilling is a machining operation that produces a hole in a solid by rotating and pressing a cutting tool (drill bit) with multiple cutting edges. In drilling, the drill bit is rotated at rates from hundreds to thousands of revolutions per minutes and pressed against the workpiece thereby resulting in cutting-off chips from the hole as it is drilled (see Figure 6.1). In engineering manufacture, drilling and related operations may be performed to either create a new hole or to enlarge existing hole in a cast metallic component or pre-machined holes in a work-part.

A notable development in drilling is the *deep-hole drilling*, which involves drilling a hole with a length to diameter ratio over 5. *Deep-hole drilling* finds applications in ship building, aerospace industry, vehicle construction, medical technology, and in hydraulic and pneumatic components manufacture. In ship building, the manufacture of the rotor shafts and components for pumps and powertrains involves deep-hole drilling operation that can be carried out with great precision, reliably, and efficiently by using deep-hole drilling machines.

Rock drilling is extensively applied in oil and gas industry; here the drilling process involves a drilling rig – an integrated system that drills wells (*e.g.*, oil wells, water wells. etc.) in the earth's subsurface (Speight, 2018). In *rock drilling*, the hole is usually made by hammering a drill bit into the hole with quickly repeated short movements. *Rock drilling* rarely leaves a clean hole suitable for long-term production since cuttings, fins, mud-weighting materials, low-gravity solids, mud filter cake, mud rings, salt accumulations, and other materials are often left behind inside a well-bore when the drilling operation is complete (Skinner, 2019).

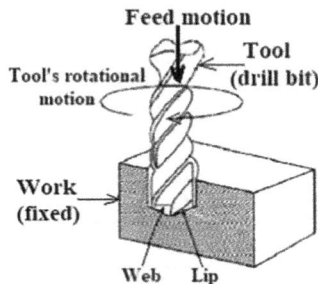

FIGURE 6.1 Drilling machining operation.

6.2 DRILLING-RELATED OPERATIONS AND THEIR APPLICATIONS

The drilling machining operation has been introduced in the preceding section. In drilling, a rotating cutting tool (drill bit) is pressed against the workpiece while ensuring the presence of lubricant; the latter is important to reduce surface roughness of the machined material and to improve its fatigue life (Sun et al., 2016). Besides the basic drilling operation, there are several drilling-related operations in practice in industry; these operations include (a) tapping, (b) reaming, (c) counter-boring, (d) counter-sinking, (e) center drilling, (f) spot facing, and the like. All drilling and the related operations involve the use of rotating cutting tools. Most of these operations follow the drilling operation *i.e.*, a hole must be made first by a drilling method, and then the hole is modified by one of the other drilling-related operations; however, center drilling and spot facing operations are exceptions to this rule. Figure 6.2 illustrates drilling and some commonly used drilling-related operations, which are explained in the following paragraph.

Tapping is performed by use of a *tap* (Figure 6.2b); it is used to provide internal threads on an existing hole. Many machine components require internal threads for inserting screws and bolts. For example, a cars-engine's oil chamber requires internal threads to introduce a threaded bolt so that contaminated oil may be drained off by opening the bolt. *Reaming* is performed by use of a *reamer* (Figure 6.2c). The purpose of reaming is three fold: (a) to enlarge an existing hole and to remove all tool marks from the pre-machined hole, (b) to provide a better tolerance on its diameter, and (c) to improve the surface finish of the hole. In particular, improved surface finish ensures a good fatigue resistance in aerospace components and the machine elements subjected to cyclic loading. *Boring* is performed by use of a special boring tool; it is used to enlarge an existing hole (Figure 6.2d). *Counter-boring* is performed by use of a *counter bore* having either straight of spiral teeth: cutting edges (Figure 6.2e). Counter-boring is used to enlarge one end of an existing hole concentric with the original hole with square bottom. The components with counter-bored holes enable

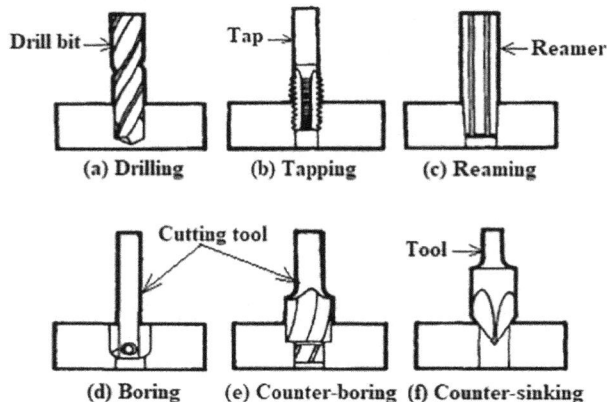

(a) Drilling   (b) Tapping   (c) Reaming

(d) Boring   (e) Counter-boring   (f) Counter-sinking

FIGURE 6.2   Drilling and some drilling-related operations

us to accommodate the heads of bolts, studs, and pins. *Countersinking* is used to produce a conical hole in a workpiece (Figure 6.2f).

6.3 DRILL BITS

6.3.1 FUNCTION, MATERIALS, AND TYPES OF DRILL BIT

Function. *Drill bits* are the cutting tools that are used to remove material to produce a hole in a solid. Although drill bits can create different kinds of holes, the holes' cross-section is generally circular. Drill bits come in many sizes and shapes and made of different materials. In order to produce a hole, the drill bit is usually attached to a drill machine, which powers them to cut through the workpiece by rotation (see Figure 6.1).

Materials. Depending on the required application, many different materials are used in making drill bits. These materials include medium-carbon steel, high-carbon steel, high-speed steel (HSS), cobalt alloy steels, tungsten carbide, cemented carbide, coated bits, and synthetic diamond. Softer medium-carbon steel bits are used for drilling wood. Although they are low-cost tools, they do not hold an edge well and require frequent sharpening. HSS bits are hard and tough and much more heat-resistant than plain carbon steel bits. They can be used to drill metal, hardwood, even at high speeds. Carbide drill bits are very hard and can drill virtually all materials, with long-lasting cutting edges. However, carbide bits are expensive. This is why they are mainly used for drill-bit tips – small pieces of carbide material fixed or brazed onto the tip of a bit made of less hard metal. Synthetic diamond is among the hardest of all tool materials and is therefore extremely resistant to wear. They are used in the automotive, aerospace, and other industries to drill carbon-fiber reinforced plastic (CFRP) and extremely hard abrasive materials.

Types of Drill Bits. Many different types of drill bits are used in machining; however, commonly used drill bits include twist bits, step bits, core bits, center bits, spade bits, and the like. Since *twist drill bit is* the most commonly used tool, it is discussed in the following sub-section.

6.3.2 TWIST DRILL BITS

Twist drills bits, also called as *twist bits*, are the most widely used of all drill bit types (Huda, 2017). They are generally made of HSS and can cut any material from wood and plastic to metal and concrete. The low-cost twist drill bits have diameters up to about ½″ (12.7 mm); they are designed for wood-working jobs. A twist bit is basically a metal rod of a specific diameter that has two, three, or four spiral flutes running most of its length (Figure 6.3). Two-flute drills are used for primary drilling jobs, whereas three-flute and four-flute twist bits are used for enlarging either an existing hole in a cast metallic component or a punched hole in a part. The section between the two flutes is called the *web,* which lies inside the drill. A point is formed by relief grinding the web to an angle of 59° from the drill's axis, which is 118° inclusive – referred to as *point angle* (see Figure 6.3a). The point angle forms a sloped cutting edge at the edge of the flute – called as the *lip* Figure 6.3b).

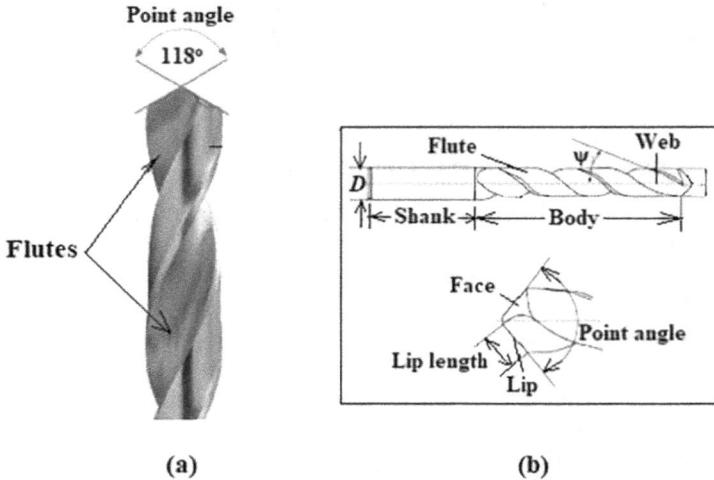

(a)   (b)

FIGURE 6.3 A twist drill bit: (a) labeled photograph of a twist bit and (b) schematic of a twist bit showing various parts of the bit. (*D* = drill diameter)

6.4 DRILLING MACHINES

6.4.1 FEATURES AND TYPES OF DRILLING MACHINES

A drilling machine, also called a *drill press*, is used to create holes into or through metals/materials. It uses a *drill bit* – a cutting tool that has cutting edges at its point (see Section 6.3). The cutting tool is held in the drill press by using a chuck or Morse taper and is rotated and fed into the work at variable speeds. A drill press operator must be competent enough to set up the machine and the work, set cutting speed and feed, and provide coolant to produce an acceptable finished product.

A large number of tools and machines have been designed for drilling of holes economically, accurately, and quickly in all types of materials (Jayendran, 1994). Drilling machines come in many shapes and sizes, from small hand-held power drills through bench-mounted models to floor-mounted machines. The major types of drilling machines include (a) upright drill press, (b) radial arm drill press, (c) multiple-spindle drilling machine, (d) gang drilling machine, (e) turret drill press, and (f) CNC drilling machines. Some commonly used drilling machines are explained in the following sub-sections.

6.4.2 UPRIGHT DRILL PRESS

The upright drill press consists of a spindle, sleeve, column, head, worktable, and base (Figure 6.4). The spindle holds the drill bit and revolves in a fixed position in a sleeve. In general, the spindle is oriented vertical, and the work is supported on a horizontal table. The sleeve may slide in its bearing in a direction parallel to its axis. When the sleeve carrying the spindle is lowered, the drill bit is fed into the work; on the other hand, when the sleeve is raised, the drill bit is withdrawn from the work.

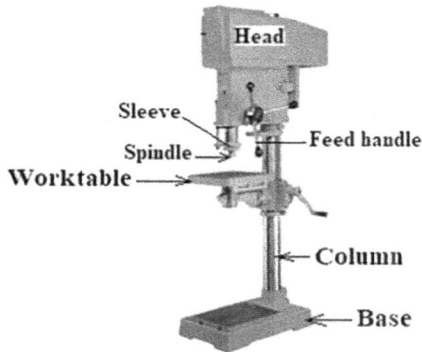

FIGURE 6.4 An upright drill press

The feed pressure applied to the sleeve either manually or by power causes the revolving drill bit to cut its way into the work. The circular column supports the head and the sleeve assembly. The head of the drill press houses the electric motor, sleeve, spindle, and the feed mechanism. The head is bolted to the column. The worktable is supported on an arm mounted to the column; it can be adjusted vertically as per height requirements of the workpiece. The entire drilling machine is supported by the base, which is bolted to the floor.

6.4.3 RADIAL ARM DRILL PRESS

The *radial arm drill press*, also called as a *radial drill press*, allows the operator to position the spindle directly over the workpiece rather than moving the workpiece to the tool. This design feature of the radial drill press enables the operator to perform machining on the parts too large to position easily. Radial drill presses offer power feed on the spindle; additionally, there is an automatic screw mechanism to raise or lower the radial arm (see Figure 6.5). The wheel head, located on the radial arm, can also be traversed along the arm, giving the machine added ease of use and versatility. Some radial arm drill presses are equipped with a tilting table; this feature allows the operator to drill intersecting or angular holes in one machine setup.

6.4.4 MULTIPLE-SPINDLE AND TURRET DRILLING MACHINES

Multiple-Spindle Drilling Machine. The multiple-spindle drilling machine, also called a *multi-spindle drill press*, is a special purpose drill press that has many spindles connected to one main work head. During the machine operation, all of the spindles are simultaneously fed into the workpiece. The multi-spindle drill press is especially useful when there are many holes located close together in a workpiece, and when there are a large number of workpieces.

Turret Drilling Machine. Turret type drilling machines are equipped with several heads that are mounted on a turret; each turret head can be equipped with a different type of cutting tool. The turret allows the operator to quickly switch to the needed tool

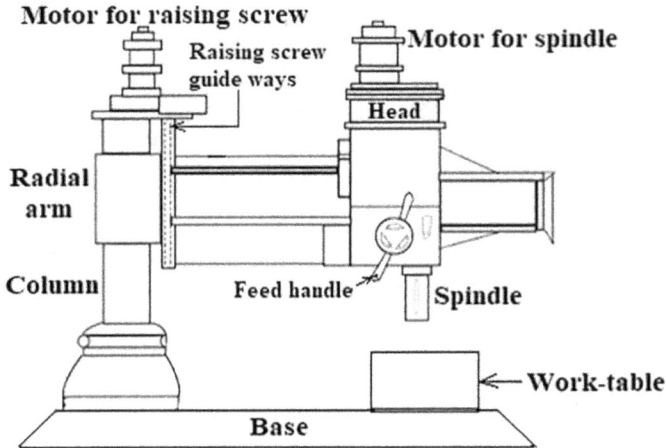

FIGURE 6.5   Schematic illustration of a radial arm drill press

into position to perform the drilling operation. In CNC turret drilling machines, the worktable can be quickly and accurately positioned for manufacturing efficiency.

6.5 ENGINEERING ANALYSIS OF DRILLING OPERATION

It has been illustrated in Sections 6.1–6.3 that drilling machining operation involves speed rotation of a drill bit and its pressing against the workpiece, thereby resulting in the cutting off chips from the hole as it is drilled (see Figures 6.1–6.3). Prior to the operation of a drill press, the machinist must select an appropriate cutting speed V based on the type of cutting tool and the work material to be machined by reference to the data in Table 6.1. Once the (surface) cutting speed for a set of work material and cutting tool has been selected, the spindle rotational speed is computed by:

$$N = \frac{V}{\pi D} \qquad (6.1)$$

where N is the spindle rotational speed, rev/min; V is the maximum (surface) cutting speed, mm/min or in/min; and D is the drill (drill bit) diameter, mm or in. (see Example 6.1).

Besides the computation for spindle rpm, it is also important to select the feed for a specified drill dimeter; the recommended feeds for various drill diameters are presented in Table 6.2.

While referring to the data in Tables 6.1 and 6.2, the machinist must select the smaller values for stronger/harder materials and the larger values for weaker/softer grades of the work materials. The significance of Tables 6.1 and 6.2 is illustrated in Example 6.2.

Once the feed has been selected by using Table 6.2, the tool feed rate can be calculated by:

TABLE 6.1
Cutting Speeds (sfpm*) for Drilling Various Work Materials Using Different Tools

No.	Material	HSS Tool (ft/min)	Carbide Tool (ft/min)
1	Aluminum and its alloys	250	625
3	Brass	250	625
4	Bronze (high tensile strength)	100	250
5	Cast iron (soft)	100	250
6	Cast Iron (medium hard)	80	200
7	Cast Iron (hard chilled)	20	50
8	Nickel alloy (Hastelloy)	20	50
9	Nickel alloy (Inconel)	25	62
10	Magnesium and its alloys	300	750
11	Monel metal	25	62
12	Plain carbon steels (0.2–0.3% C)	100	250
13	Plain carbon steels (0.4–0.5% C)	60	150
14	Steel alloy (high nickel steel)	50	125
15	Steels alloys (300–400 BHN)	30	75
16	Steel forgings and tool steels	40	100
17	Stainless steels	20–40	50–100
18	Titanium alloys	20	50

* = surface feet per minute.

TABLE 6.2
Recommended Tool Feeds for Various Drill Diameters (University of Florida, 2020)

#	Drill Diameter (in.)	Feed (in./rev)
1	Under 1/8	Up to 0.002
2	1/8–1/4	0.002–0.004
3	1/4–1/2	0.004–0.008
4	1/2–1	0.008–0.012
5	1 and over	0.012–0.020

$$f_r = f N \qquad (6.2)$$

where f_r is the feed rate, mm/min, and f is the feed, mm/rev (see Example 6.3).

The material removal rate (MRR) can be calculated by considering the time rate of volume of material removed, as follows:

$$\text{MRR} = \text{drill cross-sectional area} \times \text{feed rate} = \left(\frac{\pi}{4}D^2\right)f_r = \frac{\pi D^2 f_r}{4} \qquad (6.3)$$

where MRR is the material removal rate, mm³/min; D is the drill diameter, mm; and f_r is the feed rate, mm/min (see Example 6.4).

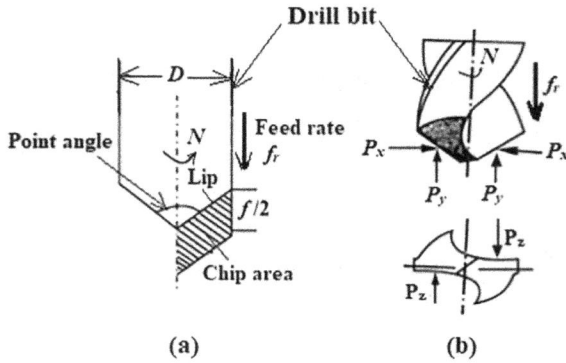

FIGURE 6.6   Drilling operation showing the chip area per lip (a) and forces in drilling (b)

In order to derive an expression for the cutting force in drilling, we refer to Figure 6.6. Figure 6.6a shows that the width of the chip per lip (w_c) is equal to half the feed *i.e.*

$$w_c = \frac{f}{2} \tag{6.4}$$

It can be easily be shown with reference to Figure 6.6 that the length of the chip per lip l_c is:

$$l_c \cong \frac{D}{2} \tag{6.5}$$

The chip area per unit lip can be computed by combining Equations 6.4 and 6.5, as follows:

$$A_c = l_c w_c = \frac{fD}{4} \tag{6.6}$$

where A_c is the chip area per lip, mm², and f is the feed, mm/rev (see Example 6.6).

Now, the cutting force can be calculated by using the following mathematical relationship:

$$F_c = K_s A_c \tag{6.7}$$

where F_c is the cutting force, N and K_s is the specific cutting force, N/mm² (K_s is defined as the cutting force needed to cut a chip area of 1 mm² having a thickness of 1 mm). The significance of Equation 6.7 is illustrated in Example 6.7.

The thrust force in drilling is a function of the force P_y, as shown in Figure 6.6b. The thrust force can be determined by using the following equation (Shaw and Oxford, 1957):

$$F_t = 10^6 \left(0.195\,H_B\,f^{0.8}D^{0.8} + 0.0022\,H_BD^2\right) \qquad (6.8)$$

where F_t is the thrust force in drilling, N; H_B is the Brinell hardness number (BHN) of the work material; f is the feed, m/rev; and D is the drill diameter, m (see Example 6.8).

The cutting torque in drilling is a function of the force P_z as indicated in Figure 6.6b. The cutting torque in drilling is calculated by:

$$Torq_c = 2 \times 10^6\,H_B^{0.7}D^2 f^{0.8} \qquad (6.9)$$

where $Torq_c$ is the torque in drilling, N · m (see Example 6.9).

The calculation for the cutting time in drilling depends on whether the hole to be drilled is a *through hole* (Figure 6.7a) or a *blind hole* (Figure 6.7b). In Figure 6.7a, h can be calculated by:

$$h = \frac{D}{2\,tan\left(\dfrac{\varphi}{2}\right)} \qquad (6.10)$$

where h is the drill point height, mm; D is the drill diameter, mm; and Φ is the drill point angle (see Example 6.10).

Once the value of h has been computed by using Equation 6.10, the cutting time for a *through hole* is calculated by dividing the length of cut by the feed rate, as follows:

$$T_c = \frac{L_c}{f_r} = \frac{h+t}{f_r} \qquad (6.11)$$

where T_c is the cutting time, min; t is the work thickness, mm; and f_r is the feed rate, mm/min (see Examples 6.11 and 6.12).

The cutting time for drilling a blind hole is calculated by (Huda, 2018):

$$T_c = \frac{d}{f_r} \qquad (6.12)$$

where d is the hole depth, mm (see Examples 6.13 and 6.14).

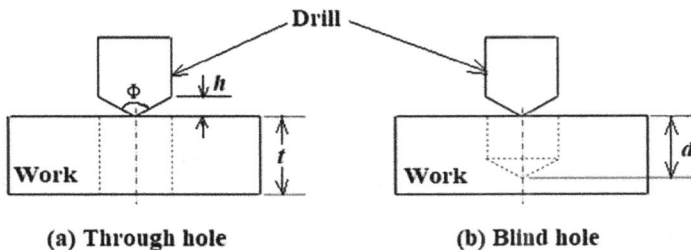

(a) Through hole   **(b) Blind hole**

FIGURE 6.7   Drilling a through hole (a) and a blind hole (b). (h = drill point height, t = work thickness, d = depth of hole, Φ = point angle)

6.6 CALCULATIONS – *WORKED EXAMPLES ON TURNING OPERATIONS*

EXAMPLE 6.1: CALCULATING THE SPINDLE RPM FOR DRILLING OPERATION

A drilling operation is performed on a Monel metal workpiece using a ½″ HSS tool of an upright drill press. Calculate the spindle rotational speed for the operation.

Solution
By reference to Table 6.1, for Monel metal, $V = 25$ ft/min $= 25 \times 12 = 300$ in./min, $D = 0.5$ in.
   By using Equation 6.1,

$$N = \frac{V}{\pi D} = \frac{300}{0.5 \times \pi} = 191 \text{rev/min}$$

The rotational speed of spindle = 191 rpm.

EXAMPLE 6.2: SELECTING THE TOOL FEED AND THE CUTTING SPEED FOR DRILLING OPERATION

A 1/4″ hole is to be drilled in a high-strength stainless steel workpiece by using a carbide tool.
   Select the feed and the cutting speed for the drilling operation.

Solution
By reference to Table 6.2, for 1/4″ drill diameter, feed = 0.004 in./rev.
   By reference to Table 6.1, Cutting speed = 50 ft/min = 600 in./min.

EXAMPLE 6.3: CALCULATING THE FEED RATE FOR DRILLING WHEN THE SPINDLE RPM IS UNKNOWN

By using the data in Example 6.2, calculate the tool feed rate for the drilling operation.

Solution
$D = ¼$ in. $= 0.25$ in. $= 0.25 \times 25.4 = 6.35$ mm, $f = 0.004$ in./rev $= 0.1016$ mm/rev, $V = 600$ in./min $= 600 \times 25.4 = 15{,}240$ mm/min.
   By using Equation 6.1,

$$N = \frac{V}{\pi D} = \frac{15{,}240}{6.35 \pi} = 763.84 \text{rev/min}$$

   By using Equation 6.2,

$$f_r = f N = 0.1016 \times 763.84 = 77.6 \, \text{mm/min}$$

EXAMPLE 6.4: CALCULATING THE MRR IN DRILLING

By using the data in Examples 6.2 and 6.3, calculate the MRR for the operation.

Solution
$D = 6.35$ mm, $f_r = 77.6$ mm/min, MRR = ?
By using Equation 6.3,

$$\text{MRR} = \frac{\pi D^2 f_r}{4} = \frac{\pi \times 6.35^2 \times 77.6}{4} = 2457.85 \, \text{mm}^3/\text{min}$$

The MRR = 2457.85 mm³/min = 2.458 cm³/min.

EXAMPLE 6.5: CALCULATING THE CUTTING POWER IN DRILLING OPERATION

By using the data in Examples 6.2–6.4, calculate the cutting power in the drilling operation.

Solution
By reference to Table 3.1, the unit power for stainless steel = $P_u = 5$ W · s/mm³, MRR = 2457.85 mm³/min = 40.96 mm³/s, $P_c = ?$
By using the modified form of Equation 3.9,

$$P_c = P_u (\text{MRR}) = 5 \times 40.96 = 204.8 \, \text{W}$$

The cutting power = 204 W.

EXAMPLE 6.6: CALCULATING THE CHIP AREA PER LIP IN DRILLING OPERATION

By using the data in Examples 6.2 and 6.3, compute the chip area per lip for the operation.

Solution
$D = 6.35$ mm, $f = 0.1016$ mm/rev
By using Equation 6.6,

$$A_c = \frac{f D}{4} = \frac{0.1016 \times 6.35}{4} = 0.1613 \, \text{mm}^2$$

The chip area per lip = 0.1613 mm².

EXAMPLE 6.7: CALCULATING THE CUTTING FORCE IN DRILLING

By using the data in Example 6.6, calculate the cutting force in the drilling operation if the specific cutting force for the work material is 2000 N/mm².

Solution
K_s = 2000 N/mm², A_c = 0.645 mm², F_c = ?
    By using Equation 6.7,

$$F_c = K_s A_c = 2000 \times 0.645 = 1290\,\text{N}$$

The cutting force = 1290 N.

EXAMPLE 6.8: CALCULATING THE THRUST FORCE IN DRILLING

A 3/4″ hole is drilled in a free-machining steel workpiece with 200 BHN. The tool feed is 0.2 mm/rev. Calculate the thrust force for the drilling operation.

Solution
D = 0.75 in. = 0.75 × 25.4 mm = 19.05 mm = 0.019 m; D^2 = 0.000361, H_B = 200, f = 0.2 mm/rev = 0.2 × 10^{-3} m/rev, $f^{0.8}$ = (0.0002)$^{0.8}$ = 1.098 × 10^{-3} = 0.001098
    By using Equation 6.8,

$$F_t = 10^6 \left(0.195\,H_B\,f^{0.8}\,D^{0.8} + 0.0022\,H_B\,D^2 \right)$$

$$= 10^6 \left[\left(0.195 \times 200 \times 0.001098 \times 0.0419 \right) + \left(0.0022 \times 200 \times 0.000361 \right) \right]$$

$$F_t = 10^6 \left(0.0001794 + 0.00015884 \right) = 3.3824 \times 10^{-4} \times 10^6 = 338.24\,\text{N}$$

The thrust force in drilling = 338.24 N.

EXAMPLE 6.9: CALCULATING THE TORQUE IN DRILLING

By using the data in Example 6.8, calculate the torque in the drilling operation.

Solution
D = 0.75 in. = 0.019 m; D^2 = 0.000361, H_B = 200, $H_B^{0.7}$ = 200$^{0.7}$ = 40.8, f = 0.2 mm/rev = 0.2 × 10^{-3} m/rev, $f^{0.8}$ = (0.0002)$^{0.8}$ = 1.098 × 10^{-3} = 0.001098
    By using Equation 6.9,

$$Torq_c = 2 \times 10^6\,H_B^{0.7}\,D^2\,f^{0.8} = 2 \times 10^6 \times 40.8 \times 0.000361 \times 0.001098$$
$$= 3.23 \times 10 = 32.3\,\text{N} \cdot \text{m}$$

The torque in the drilling operation = 32.3 N · m.

EXAMPLE 6.10: CALCULATING THE POINT HEIGHT OF A DRILL BIT

A drill bit has a point angle of 118°, and its diameter is 10 mm. Calculate the drill point height.

Solution

$\Phi = 118°$, $D = 10$ mm, $h = ?$

By using Equation 6.10,

$$h = \frac{D}{2\,tan\left(\dfrac{\varphi}{2}\right)} = \frac{10}{2\,tan\left(\dfrac{118°}{2}\right)} = 3\,\text{mm}$$

The drill point height = 3 mm.

EXAMPLE 6.11: CALCULATING THE CUTTING TIME FOR DRILLING A THROUGH HOLE

A through hole is to be drilled in a 3-cm-thick plate by using a drill bit with a point angle of 118° and drill diameter of 10 mm. The feed is 0.17 mm/rev, and the spindle rpm is 550 rpm. Calculate the cutting time for the drilling operation.

Solution

$t = 3$ cm $= 30$ mm, $h = 3$ mm (see Example 6.10), $f = 0.17$ mm/rev, $N = 550$ rev/min, $T_c = ?$

By using Equation 6.2,

$$f_r = f\,N = 0.17 \times 550 = 93.5\,\text{mm/min}$$

By using Equation 6.11,

$$T_c = \frac{h+t}{f_r} = \frac{3+30}{93.5} = 0.353\,\text{min}$$

The cutting time for drilling operation = 0.353 min = 21 s.

EXAMPLE 6.12: CALCULATING THE % INCREASE OR % DECREASE IN CUTTING TIME IN DRILLING WHEN SPINDLE RPM AND FEED ARE CHANGED TO IMPROVE SURFACE FINISH

Refer to Example 6.11. It is required to improve the surface finish of the hole by increasing the spindle rotational speed by 22% and reduce the feed by 27%. Calculate (a) the new cutting time for the drilling operation, and (b) % increase or % decrease in the cutting time.

Solution

$t = 30$ mm, $h = 3$ mm, $f = 0.17 - (27\% \times 0.17) = 0.17 - 0.046 = 0.124$ mm/rev, $N = 550 + (22\% \times 550) = 550 + 121 = 671$ rev/min, $T_c = ?$

a. By using Equation 6.2,

$$f_r = f\,N = 0.124 \times 671 = 83.2\,\text{mm/min}$$

By using Equation 6.11,

$$T_c = \frac{h+t}{f_r} = \frac{3+30}{83.2} = 0.396 \, \text{min}$$

b. % Increase in cutting time = $\left(\dfrac{\mathbf{0.396 - 0.352}}{\mathbf{0.352}} \right) \times 100 = 12.68.$

or Increase in the cutting time = 12.68%.

EXAMPLE 6.13: CALCULATING THE DEPTH OF A BLIND HOLE

It took 1.5 min to drill a blind hole to a certain depth by performing the drilling operation with spindle rotational speed of 300 rpm and feed of 0.21 mm/rev. What is the depth of the hole?

Solution
$T_c = 1.5$ min, $N = 300$ rev/min, $f = 0.21$ mm/rev, $d = ?$
By using Equation 6.2,

$$f_r = f \, N = 0.21 \times 300 = 63 \, \text{mm/min}$$

By using the modified form of Equation 6.12,

$$d = T_c \, f_r = 1.5 \times 63 = 94.5 \, \text{mm} = 9.45 \, \text{cm}$$

The hole depth = 9.45 cm.

EXAMPLE 6.14: COMPUTING THE CUTTING TIME FOR DRILLING WHEN THE FEED RATE, SPINDLE RPM, AND THE FEED ARE NOT GIVEN

An 8-mm-diameter drill bit made of cemented carbide is used to drill a 10-mm-deep blind hole in a titanium alloy plate. Calculate the cutting time to perform the drilling operation.
Hint: Take average value of the recommended feed in Table 6.2.

Solution
By reference to Table 6.1, the cutting speed for titanium alloy with carbide tool = V = 50 ft/min

$$V = 50 \times 12 \times 25.4 \, \text{mm/min} = 15240 \, \text{mm/min}$$

By reference to Table 6.2, the average feed for 8-mm-diameter or 0.31″-diameter drill is 0.006 in./rev *i.e.*, $f = 0.006$ in./rev = 0.15 mm/rev, $V = 15,240$ mm/min, $f = 0.15$ mm/rev, $d = 10$ mm, $D = 8$ mm, $T_c = ?$
By using Equation 6.1,

$$N = \frac{V}{\pi D} = \frac{15,240}{8\pi} = 606.3 \, \text{rev/min}$$

By using Equation 6.2,

$$f_r = fN = 0.15 \times 606.3 = 90.94 \, \text{mm/min}$$

By using Equation 6.12,

$$T_c = \frac{d}{f_r} = \frac{10}{90.94} = 0.11 \, \text{min}$$

The cutting time = 0.11 min = 6.6 s.

QUESTIONS AND PROBLEMS

6.1 Underline the most appropriate answers for the following statements.
  a. Which drilling-related operation is performed to produce internal screw threads?
     (i) Tapping, (ii) reaming, (ii) counter-boring, (iv) counter-sinking
  b. Which drilling-related operation is performed to produce a conical hole in a workpiece?
     (i) Tapping, (ii) reaming, (ii) counter-boring, (iv) counter-sinking
  c. Which operation is used to enlarge one end of an existing hole concentric with the original hole with square bottom?
     (i) Tapping, (ii) reaming, (ii) counter-boring, (iv) counter-sinking
  d. Which operation is used to enlarge an existing hole and to remove tool marks from it?
     (i) Tapping, (ii) reaming, (ii) counter-boring, (iv) counter-sinking
  e. Which type of drilling machine allows the operator to drill angular holes in one set-up?
     (i) Upright drill press, (ii) radial drill press, (iii) multiple-spindle press, (iv) turret drill press
  f. Which type of drilling machine allows the operator to quickly switch to the needed tool into position to perform the drilling operation?
     (i) Upright drill press, (ii) radial drill press, (iii) multiple-spindle press, (iv) turret drill press
6.2 a. Define deep hole drawing, and draw a labelled sketch of drilling operation.
   b. Briefly explain the industrial application of drilling machining operation.
6.3 Draw sketches for six (6) drilling-related operations, and briefly explain them.
6.4 a. List the drill bit materials and give their applications.
   b. Draw a sketch of a twist drill showing its main parts.
6.5 a. Draw a labeled sketch of an upright drill press.
   b. What are the functions of various parts of an upright drill press?

6.6   Compare the capabilities of at least four types of drilling machines.

P6.7   A 1/8″ hole is to be drilled in a forged steel workpiece by using an HSS tool. Select the following cutting conditions for the operation: (a) the cutting speed and (b) the feed.

P6.8   A 1/2″ hole is to be drilled in an aluminum workpiece by using an HSS tool. Calculate the following machining parameters for the drilling operation: (a) feed rate, (b) MRR, (c) cutting power, and (d) chip area per lip.

P6.9   The chip area per lip in a drilling operation is 0.82 mm². Calculate the cutting force in the drilling operation if the specific cutting force for the work material is 1880 N/mm².

P6.10 A 1/4″-hole is drilled in a brass workpiece with 60 BHN. Calculate the thrust force and the cutting torque for the drilling operation.

P6.11 A drill bit has a point angle = 118° and diameter = 16 mm. Compute the drill point height.

P6.12 A through hole is drilled in a 2-cm-thick plate by using a drill bit with a point angle of 118° and drill diameter of 8 mm. The feed is 0.15 mm/rev, and the spindle rpm is 450. Calculate the cutting time for the drilling operation. If the spindle rpm is increased by 15% and the feed is reduced by 25%, calculate the new cutting time. Also calculate the % increase or % decrease in the cutting time due to the change in the cutting conditions.

P6.13 It took 1.1 min to drill a blind hole to a certain depth by performing drilling with spindle rotational speed of 350 rpm and feed of 0.2 mm/rev. Compute the depth of the hole.

REFERENCES

Huda, Z. (2018), *Manufacturing: Mathematical Models, Problems, and Solution*, CRC Press, Boca Raton, FL.

Huda, Z. (2017), *Materials Processing for Engineering Manufacture*, Trans Tech Publications, Pfaffikon, Switzerland.

Jayendran A. (1994) Drills and Drilling Machines. In: *Englisch für Maschinenbauer. Viewegs Fachbücher der Technik*. Vieweg+Teubner Verlag, Springer Fachmedien Wiesbaden, Germany.

Shaw, M.C., Oxford, C.J. Jr. (1957), On the drilling of metals 2 – The torque and thrust in drilling, *Transactions on ASME*, 79, 191–231.

Speight, J.G. (2018), *Formulas and Calculations for Drilling Operations*, 2nd Edition, John Wiley & Sons, New York.

Sun, D., Keys, D., Jin, Y., Malinov, S., Zhao, Q., Qin, X. (2016), Hole making and its impact on the fatigue response of T-6Al-4Al alloy, *Procedia CRIP*, 56, 289–292.

Skinner, L. (2019), *Hydraulic Rig Technology and Operations*, Gulf Professional Publishing Elsevier Science Publications, Alpharetta.

University of Florida, EML2322L – MAE Design and Manufacturing Laboratory (2020), Drilling speeds and feeds, accessed on April 19, 2020, available Online: https://mae.ufl.edu/designlab/Lab%20Assignments/EML2322LDrilling%20and%20Milling%20Speeds%20and%20Feeds.pdf

7 Milling Operations and Machines

7.1 MILLING AND ITS INDUSTRIAL IMPORTANCE

Milling is a machining operation that involves the removal of material from a work-part by use of a rotating cylindrical tool with multiple cutting edges – a *milling cutter*. In *milling*, the work has a feed motion, whereas the cutter has *speed* motion (Bray, 2004). Milling machines are primarily used for machining both metallic and non-metallic solids (e.g., metal, wood, plastics, etc.). These machines find wide-spread applications in diversified engineering industries. For example, the aerospace industry relies on machinery with a high level of precision and accuracy to manufacture aircraft parts that fit the exact specifications of aeronautical design. Biomedical manufacturing companies produce life-saving devices that are used in hospitals and clinics across the world. The manufacture of these life-saving devices and many other engineering components involve a great deal of milling operations. In particular, gears and many machine elements can be manufactured by milling operations (see Chapter 10).

7.2 FORMS OF MILLING – *PERIPHERAL MILLING AND FACE MILLING*

There are two main forms of milling operation: (a) peripheral milling and (b) face milling. In *peripheral milling*, the cutter axis is oriented parallel to surface being machined, and the cutting edges are provided on outside periphery of the cutter (see Figure 7.1a). It has experimentally been shown that successful surface finishing of hardened steel can be achieved by peripheral milling (Hayasaka et al., 2017). In *face milling*, the cutter axis is normal to the surface being milled, and cutting edges are

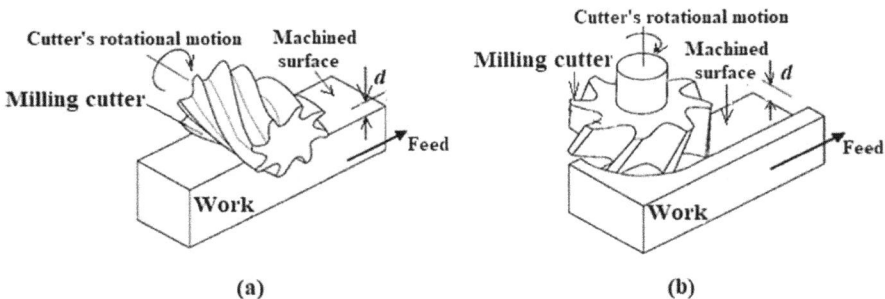

FIGURE 7.1 Two forms of milling: (a) peripheral milling and (b) face milling. (d = depth of cut)

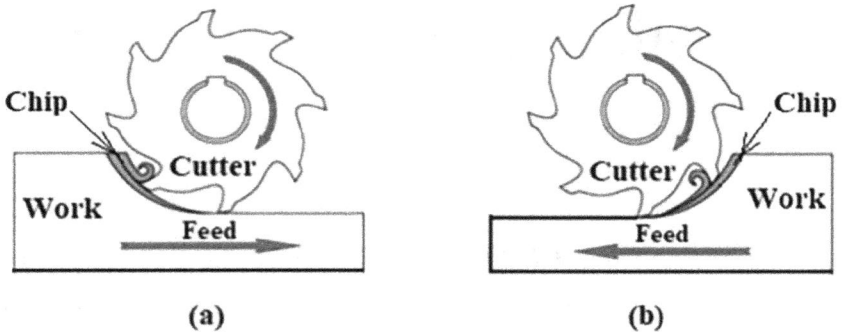

FIGURE 7.2   Methods of milling: (a) up milling and (b) down milling

provided not only on the outside periphery of the cutter but also on its end (see Figure 7.1b).

7.3 METHODS OF MILLING – *UP MILLING AND DOWN MILLING*

The milling operation may be carried out by two methods: (1) up milling and (2) down milling. In *up milling*, the cutter rotates in a direction opposite to the feed (Figure 7.2a). *Up milling* is used for rough and deep cutting. In *down milling*, the cutter rotation is along the direction of feed (Figure 7.2b). *Down milling* is generally used to achieve a better surface finish. The different features of up milling and down milling are presented in Table 7.1.

7.4 MILLING MACHINES – *CUTTERS AND TYPES OF MILLING MACHINES*

7.4.1 MILLING CUTTERS

Milling cutters are made of either high-speed steel (HSS) or cemented carbide. Depending upon the orientation of teeth, a milling cutter may be either helical or

TABLE 7.1
Distinction between Up Milling and Down Milling

#	Up Milling	Down Milling
1.	The cutter's rotation is opposite to the direction of feed motion	The cutter's rotation is along the direction of feed motion
2.	It is used for rough cutting with larger depth of cut (d)	It is used for finish machining with shorter depth of cut (d)
3.	The chip formed is longer	The chip formed is shorter
4.	The cutting tool life is shorter	The cutting tool life is longer
5	Greater clamping force is required to hold the work-part	Greater clamping force is required to hold the work-part.

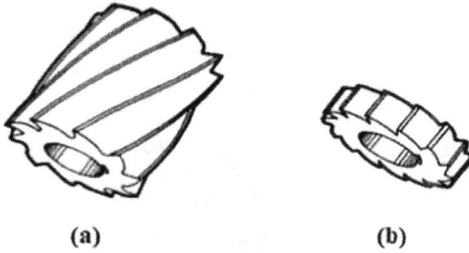

FIGURE 7.3 Milling cutters: (a) helical cutter and (b) plain cutter

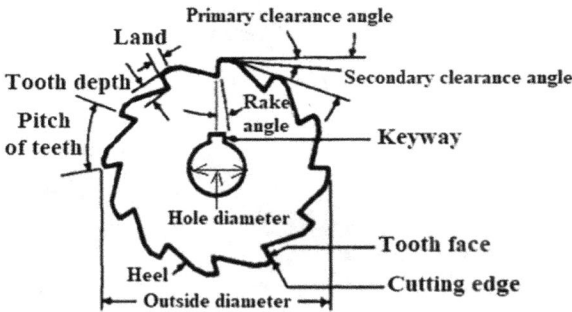

FIGURE 7.4 Milling cutter nomenclature

plain milling cutter (see Figure 7.3). The nomenclature of a milling cutter is presented in Figure 7.4. The geometrical design of a cutter plays an important role in reducing the cutting time in milling (see Section 7.6.3).

7.4.2 Milling Machines and Their Types

The technological and industrial applications of milling machines have been emphasized in Section 7.1. Milling machines are designed and constructed in different types so as to enable a machinist to perform a variety of milling operations (see Section 7.5). There are three basic types of milling machines: (1) knee-and-column-type milling machine, (2) universal horizontal milling machine, and (3) ram-type milling machines.

7.4.2.1 Knee-and-Column-Type Milling Machines

A *knee-and-column-type milling machine* is characterized by a vertically adjustable worktable resting on a saddle which is supported by a knee. The knee is a large casting that is oriented vertically on the machine's column and can be clamped rigidly to the column in a position where the milling head and the spindle are properly adjusted vertically for operation (see Figure 7.5). There are two types of *knee-and-column-type milling machine*: (a) vertical-spindle milling machines and (b) horizontal-spindle milling machines (Walker and Dixon, 2013). In a *vertical spindle milling machine*,

Milling head

Ram

Spindle

Table

Saddle

Column

Knee

Base

(a)

Arbor

Cutter

Table

Saddle

Knee

Column

Controls

Base

(b)

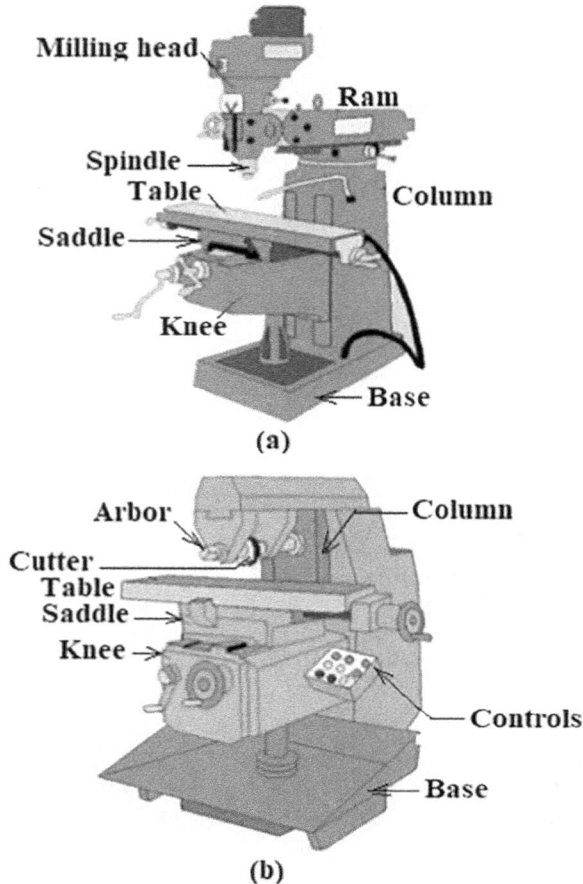

FIGURE 7.5 Milling machines: (a) vertical milling machine and (b) horizontal milling machine

the spindle is oriented vertically, parallel to the column face, and mounted in a sliding head that can be fed up and down by hand or power (Figure 7.5a). Modern vertical-spindle milling machines are designed so as to swivel the entire head; this design feature permits working on angular surfaces of workpieces. The *horizontal-spindle milling machine*'s column contains the drive motor and gearing and a fixed positioned horizontal spindle (Figure 7.5b). In a horizontal-spindle milling machine, an adjustable overhead arm containing one or more arbor supports projects forward from the top of the column. The arm and arbor supports are used to stabilize long arbors.

7.4.2.2 Universal Horizontal Milling Machines

The main difference between a *universal horizontal milling machine* and a *horizontal-spindle knee-and-column-type milling machine* is the addition of a table swivel housing between the table and the saddle of the *universal machine*. This additional feature in the universal horizontal milling machines permits the table to swing up to

45° in either direction for angular and helical milling operations. The universal horizontal milling machine can be fitted with various attachments such as the indexing head (see Section 7.5), rotary table, slotting and rack cutting attachments, and various special fixtures.

7.4.2.3 Ram-Type Milling Machines

In a *ram-type milling machine*, the spindle is mounted to a movable housing on the column so as to permit positioning of the milling cutter forward or rearward in a horizontal plane. In general, there are two types of ram-type milling machines: (a) universal ram-type milling machines, and (b) swivel cutter head ram-type milling machines. The universal ram-type milling machine is similar to the universal horizontal milling machine, and the difference is that in the universal ram-type milling machine, the spindle is mounted on a ram. In a swivel-cutter-head ram-type milling machine, the cutter head containing the milling machine spindle is attached to the ram. The cutter head can be either swiveled from a vertical spindle position to a horizontal-spindle position or it can be fixed at any desired angular position between vertical and horizontal.

7.5 MILLING OPERATIONS

7.5.1 Operations on a Vertical-Spindle Knee-and-Column-Type Milling Machine

A number of milling operations can be performed on a *vertical-spindle knee-and-column type milling machine*; these operations include face milling, end milling, pocket milling, key milling, profile milling, and the like (Hall, 2013). Some of the vertical-spindle milling machine operations are described in the following paragraphs.

> *Face Milling*. In *face milling*, the cutter axis is normal to the surface being milled. In a conventional face milling operation, the cutter overhangs work-part on both sides *i.e.*, the cutter diameter is greater than the work width (Figure 7.6a).
> *End Milling*. In *end milling*, the cutter diameter is less than the work width. The *end milling* operation can be used for slot milling (Figure 7.6b) as well as for pocket milling.
> *Pocket Milling*. In *pocket milling*, the cutter (end mill) removes material from the work thereby creating a pocket in the form of a rectangular or circular cavity (see Figure 7.6c).
> *Profile milling*. It refers to milling the outside periphery of a flat part.

7.5.2 Operations on a Horizontal-Spindle Knee-and-Column-Type Milling Machine

The horizontal-spindle milling machine operations include slab milling, slot milling, slitting, side milling, straddle milling, form milling (*e.g.*, milling of a gear), and the like. Some of these milling operations are explained in the following paragraphs.

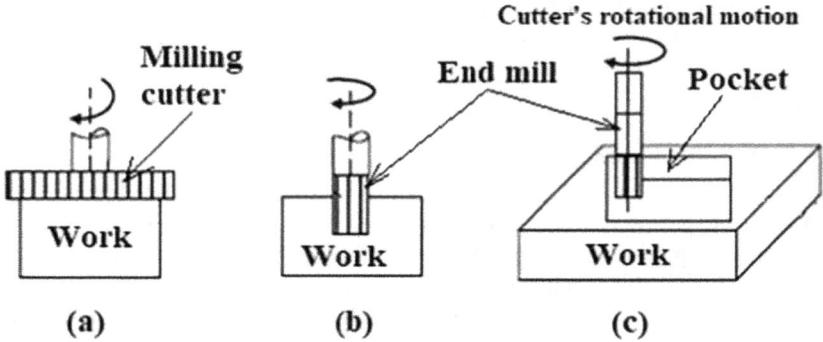

FIGURE 7.6 Some vertical-spindle milling machine operations: (a) face milling, (b) end milling, and (c) pocket milling

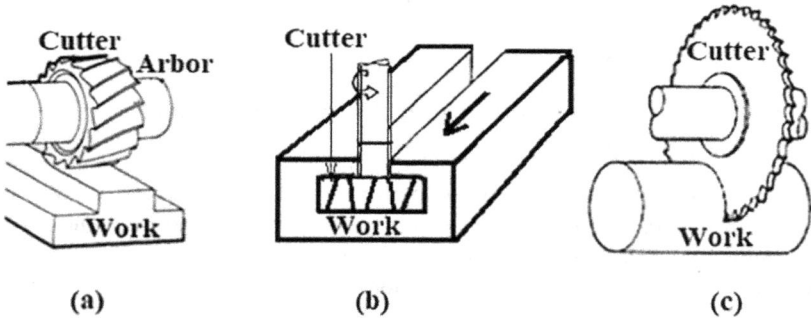

FIGURE 7.7 Some horizontal-spindle milling machine operations: (a) slab milling, (b) slot milling, and (c) slitting.

Slab Milling. It is the basic form of peripheral milling in which the cutter width exceeds the work-part width (see Figure 7.7a).

Slot Milling or Slotting. In *slotting*, a slot is milled by use of a cutter having width smaller than the work-part width (see Figure 7.7b).

Saw Milling or Slitting. It involves cutting off a work into two parts by use of a very thin cutter (see Figure 7.7c).

Side Milling. It involves milling of the side of the work by using a horizontally oriented cutter.

Straddle Milling. It is the modified form of side milling in which both sides of the work are machined (milled).

Form Milling. In form milling, the cutter teeth have the desired shape and geometry as the work. Gears can be manufactured by form milling.

TABLE 7.2

Milling cutting speeds and feed per tooth for various work materials using HSS cutter*◆

#	Work Material	Cutting Speed (m/min)	Feed per Tooth (mm/tooth)
1	Aluminum and its alloys	150–300	0.102–0.305
2	Brass and bronze	20–35	0.050–0.305
3	Copper	35	0.102–0.305
4	Gray cast iron	25	0.050–0.076
5	White cast iron	15	0.025–0.050
6	Machinable steel	21–30	0.102–0.152
7	Tool steel	18–20	0.050–0.102

*Cutting speeds may be doubled for carbide cutters.

◆**The data refers to milling cutter with outside diameters in the range of 12–40 mm.**

7.6 ENGINEERING ANALYSIS OF MILLING

7.6.1 CUTTING SPEED AND FEED IN MILLING

The cutting speeds and feeds for milling of various engineering materials have been recommended, and a partial list of milling speeds data is presented in Table 7.2.

It may be noted that the lower limit for the ranges in Table 7.2 refers to harder grades, whereas the upper limit refers to softer grades of the specified material. The cutting speeds listed in Table 7.2 correspond to the small-sized cutter with outside diameters in the range of 12–40 mm; for larger diameter cutters, significantly higher cutting speeds are applicable (see Example 7.10).

7.6.2 CUTTER'S ROTATIONAL SPEED, TABLE FEED RATE, AND MRR

Once a cutting speed (v) for a set of work-material tool has been selected, the cutter's rotational speed (N) can be calculated by the formula (Huda, 2018):

$$N = \frac{v}{\pi D} \tag{7.1}$$

where N is the cutter's rotational speed, rev/min (or spindle rpm), v is the cutting speed, mm/min, and D is the cutter's outside diameter, mm. Accordingly, based on the computed N value, the machinist must set the spindle rpm on the milling machine (see Examples 7.1 and 7.2).

The table feed rate can be calculated by the following mathematical relationship:

$$f_r = f_t N n_t \tag{7.2}$$

where f_r is the feed rate, mm/min; f_t is the feed per tooth, mm/tooth; and n_t is the number of effective teeth on the cutter (see Example 7.3). The various geometric parameters in a milling operation are shown in Figure 7.8.

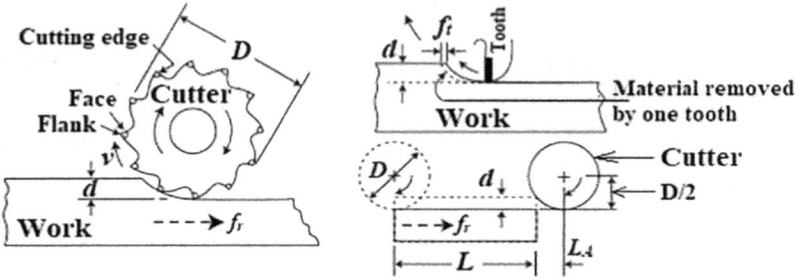

FIGURE 7.8 Geometric parameters in peripheral or slab milling (v = cutter's surface speed, d = depth of cut, f_r = feed rate, f_t = feed per tooth)

The MRR can be computed by:

$$MRR = w\, d\, f_r \qquad (7.3)$$

where MRR is material removal rate, mm³/min; w is width of the workpiece, mm; and d is depth of cut, mm (see Examples 7.4 and 7.9).

7.6.3 CUTTING TIME IN MILLING

The cutting time for one pass in *peripheral milling* can be found as follows:

$$T_c = \frac{L + L_A}{f_r} \qquad (7.4)$$

where T_c is the cutting time, min; L is the length of cut, mm; L_A is the approach distance, mm; and f_r is the feed rate, mm/min (see Figure 7.8).

The approach distance (L_A) can be found as follows:

$$L_A = \sqrt{d(D-d)} \qquad (7.5)$$

The significance of Equations 7.4 and 7.5 is illustrated in Example 7.5.

The cutting time for one pass in *face milling* can be calculated by the formula:

$$T_c = \frac{L + L_o + L_A}{f_r} \qquad (7.6)$$

where L is the length of cut, mm; L_A is the approach length, mm; L_o is the length of "over travel", mm; and f_r is the feed rate, mm/min (see Figure 7.9).

Figure 7.9 indicates that the approach length, L_A, depends on cutter's milling position *i.e.* whether the cutter is centered over the workpiece or offset to one side of the work.

When the cutter is centered over the work ($w \geq D/2$) [Figure 7.9(b)], L_A is given by:

Top views

Cutter's rotational motion

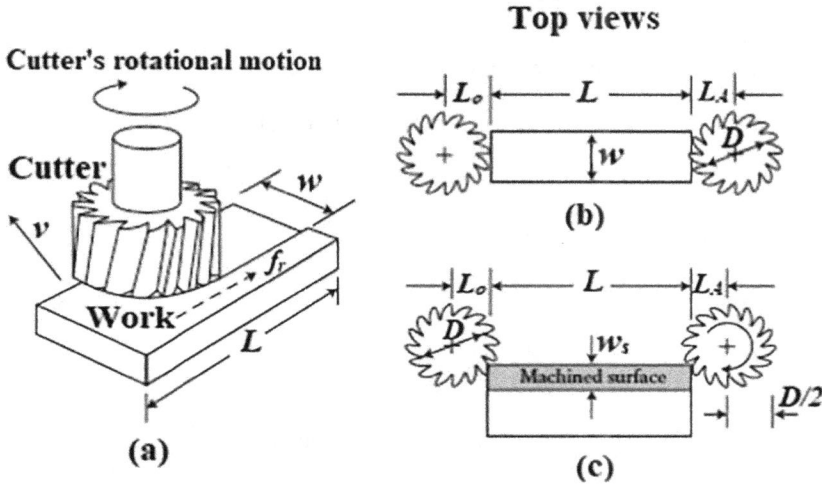

(a)

(b)

(c)

FIGURE 7.9   Geometric parameters in *face milling*: (a) face milling operation, (b) the cutter is centered over the work ($w > D/2$), and (c) the cutter is offset ($w_s < D/2$)

$$L_A = L_o = \frac{D}{2} \qquad (7.7)$$

When the cutter is offset to one side of the work ($w_s < D/2$) [Figure 7.9(c)], L_A is given by:

$$L_A = L_o = \sqrt{w_s (D - w_s)} \qquad (7.8)$$

where w_s is the width of swath cut. The significances of Equations 7.6–7.8 and Figure 7.9 are illustrated in Examples 7.6–7.8.

In industrial practice, reduction in cutting time is important for reducing labor cost and improving productivity. Equations 7.4–7.6 indicate that the cutting time in milling varies inversely as the table feed rate; the latter depends on spindle rpm and feed per tooth. It has been experimentally shown that improvement in milling cutter's geometric design plays an important role in controlling feed per tooth, thereby reduction in milling time (Vavruska et al., 2018).

7.6.4 CUTTING POWER AND TORQUE IN MILLING

The cutting power in milling depends on the specific cutting force as well as other cutting conditions. The specific cutting force, F_{sc}, may be defined as the *force in the cutting* direction required to *cut* a chip area of 1 mm^2 that has a thickness of 1 mm. The F_{sc} data for selected materials for specified feeds per tooth are given in Table 7.3.

It must be noted in Table 7.3 that the lower limits in the F_{sc} range refer to the softer grade, whereas the upper limits refer to the harder/stronger grades of the material. The significance of Table 7.3 is illustrated in Example 7.10.

TABLE 7.3

The F_{sc} Data for Selected Materials for Specified Feeds per Tooth

Work Material	Specific Cutting Force F_{sc} (MPa)		
	0.1 mm/tooth	0.3 mm/tooth	0.5 mm/tooth
Mild steel	2200	1820	1580
Medium carbon steel	1980	1730	1570
High carbon steel	2520	2040	1740
Tool steel	1980–2030	1730–1750	1590
Cr–Mn alloy steel	2300–2750	1880–2060	1660–1780
Cr–Mo alloy steel	2180–2540	1860–2140	1670–1800
Ni–Cr–Mo steel	2000–2100	1680–1760	1500–1530
Cast irons	2180–2800	1750–2320	1470–2040
Grey cast iron	1750	1240	970
Brass	1150	800	630
Aluminum alloys	580–850	400–800	320–700

The cutting power in milling is given by:

$$P_c = \frac{d \cdot w \cdot f_r \cdot F_{sc}}{60 \times 10^6 \, \eta} \tag{7.9}$$

where P_c is the cutting power, kW; d is the depth of cut, mm; w is the width of cut, mm; f_r is the table feed rate, mm/min; F_{sc} is the specific cutting force, N/mm² or MPa; and η is the machine's efficiency (see Example 7.11). It must be noted that Equation 7.9 holds good for simple rectangular workpieces; for milling involving complex geometry workpieces (e.g., gear milling), modified expressions for cutting power are required (see Section 10.2).

The torque T (in N-m) in milling is given by:

$$T = \frac{P_c}{2\pi N} \tag{7.10}$$

where P_c is the cutting power, W and N is spindle rotational speed, rev/s (see Example 7.12).

7.7 INDEXING IN MILLING

7.7.1 Indexing and Indexing Head

One of the important functions of a milling machine is to cut slots, grooves, or teeth that are to be equally spaced around the circumference of a blank. The process of dividing a part (blank) into equal spaces is called *indexing*. In particular, gear cutting involves a great deal of indexing. For indexing, the blank (workpiece) is held and rotated by the exact amount for each groove or teeth (to be cut) by use of a *dividing head* or *indexing head* (see Figure 7.10).

FIGURE 7.10   An *indexing head* mounted on the table of a small milling machine

7.7.2 Methods of Indexing

There are a number of methods of indexing; however, three indexing methods are generally practiced in milling: (a) direct indexing, (b) simple or plain indexing, and (c) differential indexing.

7.7.2.1 Direct Indexing

In *direct indexing*, a *driving plate* or *direct indexing plate* is permanently attached to the spindle. The driving plate has 24 holes, and the periphery of job can be directly divided into 2, 3, 4, 6, 8, and 12 equal parts. In milling by direct indexing, an index pin is pulled out of a hole, and the work and the driving plate are rotated according to the desired number of holes; then the pin is engaged *i.e.*, the work is locked into the place before the milling cutter is applied. *Direct indexing* is most commonly used for indexing fixtures.

7.7.2.2 Simple or Plain Indexing

In *simple indexing*, different index plates with varying number of holes are used to increase the range of indexing. The index plate is fixed in position by a pin called *index pin*. The spindle is then rotated by rotating the *index crank* which is keyed to the worm-shaft (see Figure 7.11).

Most *indexing heads* operate at a 40:1 ratio *i.e.*, 40 turns of the *index crank* results in 1 revolution of the spindle or blank. Thus, the following mathematical relation holds good:

$$\text{Number of turns of index crank required for each division of work} = \frac{40}{Z} \quad (7.11)$$

where Z is the number of divisions required in the work (blank). In *simple indexing*, the index plates are provided in sets of several plates. The Index plates with circles of holes patented by the *Brown and Sharpe Company* are presented in Table 7.4.

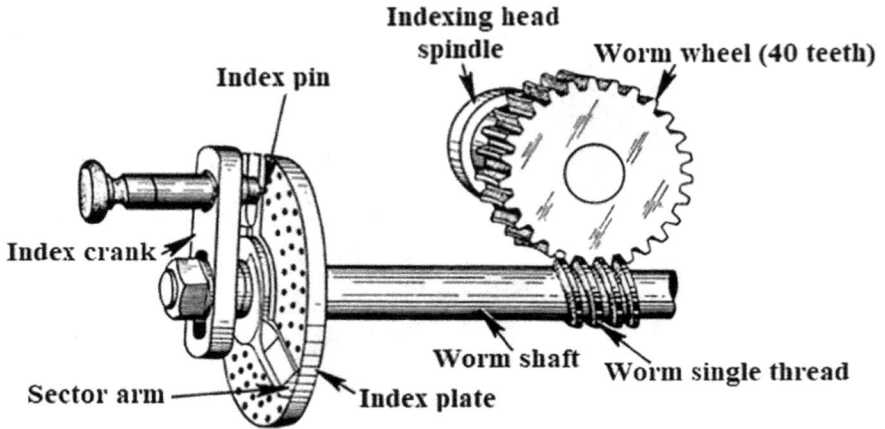

FIGURE 7.11   Simple indexing mechanism of an indexing head

TABLE 7.4
The index plates patented by the *Brown and Sharpe Company*

Plate No.	The Number of Holes in the Index Plate
1	15, 16, 17, 18, 19, 20 holes
2	21, 23, 27, 29, 31, 33 holes
3	37, 39, 41, 43, 47, 49 holes

The significance of Equation 7.11 and Table 7.4 is illustrated in Examples 7.13–7.16.

7.7.2.3 Differential Indexing

Differential indexing enables a wide range of divisions to be indexed, which cannot be accomplished by simple indexing. For differential indexing, a certain set of the change gears is incorporated extra between the indexing plate and spindle of dividing head; the latter are provided with a standard set of gears (see Figure 7.12). Thus it is possible to index all numbers from 1 to 382. As the index crank is turned, the spindle is rotated through the worm and the wheel (dividing head); this action results in the rotation of the index plate either in the same or opposite direction to that of the crank.

The following rules are applicable for differential indexing.

1. A number nearest to Z is chosen preferably from the available change gears; this nearest number is designated as Z′. This technique enables us to find a circular row with the number of holes on the available index plate thereby allowing a division into Z′ parts.

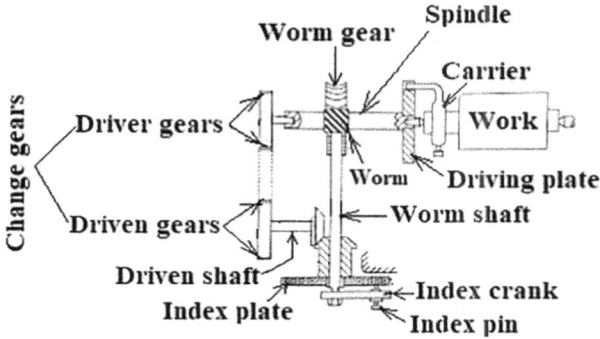

FIGURE 7.12   Differential indexing mechanism

2. The number of revolutions of the index crank is found by:

$$\text{Number of revolutions of index crank required for each division} = \frac{Z_o}{Z'} = \frac{40}{Z'} \quad (7.12)$$

3. The ratio of the differential change gears is determined by:

$$\frac{Z_1}{Z_2} \cdot \frac{Z_3}{Z_4} = \frac{Z_o(Z' - Z)}{Z'} \quad (7.13)$$

where Z_1, Z_2, Z_3, Z_4 are the number of teeth in the available change gears (Z_1 and Z_3 are the number of teeth in the driver gears whereas Z_2 and Z_4 are the number of teeth in the driven gears) (see Figure 7.12)

4. If the change gears ratio is positive, the index plate rotation is in the same direction as that of the crank, and an idle gear is used.
5. If the ratio is negative, the index plate rotation is in a direction opposite to that of the crank, and two idle gears must be used.

The significance of the differential indexing rules is illustrated in Examples 7.17–7.20.

7.8 CALCULATIONS – WORKED EXAMPLES ON MILLING

EXAMPLE 7.1: SELECTING THE CUTTING SPEED AND CALCULATING SPINDLE RPM

A brass work-part is to be machined by using an HSS milling cutter with 35 mm outside diameter. (a) Select the cutting speed in mm/min, (b) calculate the cutter rotational speed, and (c) select the spindle rpm if the available spindle rpm in the machine are 100, 200, 300, 400.

Solution

    a. $D = 35$ mm, $v = 25$ m/min $= 25,000$ mm/min (see Table 7.2 for brass).

    b. By using Equation 7.1,

$$N = \frac{v}{\pi D} = \frac{25,000}{\pi \times 35} = 227.3 \,\text{rev/min}$$

    c. The nearest and safe available spindle rpm is 200.

EXAMPLE 7.2: CALCULATING THE SPINDLE RPM AFTER CHANGING THE MILLING CUTTER

Repeat Example 7.1 if the milling cutter is made of cemented carbide.

    a. For carbide cutting tool, $v = 2 \times 25$ m/min $= 50,000$ mm/min (see Table 7.2)

    b. By using Equation 7.1,

$$N = \frac{v}{\pi D} = \frac{50,000}{\pi \times 35} = 455 \,\text{rev/min}$$

    c. The nearest available spindle rpm is 400.

EXAMPLE 7.3: SELECTING THE FEED PER TOOTH AND CALCULATING THE FEED RATE

A tool steel work-part (280 mm × 70 mm × 18 mm) is to be machined by peripheral milling using a carbide cutter rotating at 170 rpm. The cutter details are shown in Figure 7.8. Select the feed per tooth and calculate the feed rate.

Solution

By reference to Table 7.2 for the tool steel work and carbide cutting tool, $f_t = 0.08$ mm/tooth

$$N = 170 \,\text{rpm}, n_t = 12 \quad \left(\text{see Figure 7.8}\right)$$

By using Equation 7.2,

$$f_r = f_t \, N \, n_t = 0.08 \times 170 \times 12 = 163.2 \,\text{mm/min}$$

The feed rate $= f_r = 163.2$ mm/min.

EXAMPLE 7.4: CALCULATING THE MRR IN PERIPHERAL MILLING

By using the data in Example 7.3, calculate the MRR if it is required to remove 6 mm from the top surface of the workpiece.

Solution

From Example 7.3, $w = 70$ mm, $f_r = 163.2$ mm/min, $d = 6$ mm

By using Equation 7.3,

$$\text{MRR} = w\, d\, f_r = 70 \times 6 \times 163.2 = 68,544 \,\text{mm}^3/\text{min}$$

The MRR = 68,544 mm³/min.

EXAMPLE 7.5: CALCULATING THE CUTTING TIME IN PERIPHERAL MILLING

By using the data in Example 7.4, calculate the cutting time for milling one pass by using the cutter with an outside diameter of 30 mm.

Solution

From Example 7.4, $L = 280$ mm, $D = 30$ mm, $f_r = 163.2$ mm/min, $d = 6$ mm

By using Equation 7.5,

$$L_A = \sqrt{d(D-d)} = \sqrt{6(30-6)} = \sqrt{144} = 12\,\text{mm}$$

By using Equation 7.4,

$$T_c = \frac{L+L_A}{f_r} = \frac{280+12}{163.2} = 1.8\,\text{min}$$

The cutting time for milling one pass = 1.8 min.

EXAMPLE 7.6: CALCULATING THE CUTTING TIME AND MRR IN FACE MILLING WHEN THE CUTTER IS CENTERED OVER THE WORK

It is desired to remove 5 mm from the top surface of a copper rectangular workpiece (320 mm × 135 mm) in one pass by a face milling operation. The cutter is centered over the work. The HSS cutter has 10 teeth, and its outer diameter is 40 mm. Calculate (a) the cutting time to make the single pass and (b) the metal removal rate.

Solution

$d = 5$ mm, $L = 320$ mm, $w = 135$ mm, $D = 40$ mm, $n_t = 10$.

By reference to Table 7.2, $v = 35$ m/min = 35,000 mm/min, $f_t = 0.2$ mm/tooth

By using Equations 7.1, 7.2, and 7.7,

$$N = \frac{v}{\pi\, D} = \frac{35,000}{\pi \times 40} = 278.5\,\text{rev/min}$$

$$f_r = f_t\, N\, n_t = 0.2 \times 278.5 \times 10 = 557\,\text{mm/min}$$

$$L_A = L_o = \frac{D}{2} = \frac{40}{2} = 20\,\text{mm}$$

a. By using Equation 7.6,

$$T_c = \frac{L + L_o + L_A}{f_r} = \frac{320 + 20 + 20}{557} = 0.65\,\text{min} = 38.8\,\text{s}$$

b. By using Equation 7.3,

$$\text{MRR} = w\,d\,f_r = 135 \times 5 \times 557 = 375{,}975\,\text{mm}^3/\text{min}$$

The cutting time for one milling pass is 38.8 s, and the metal removal rate is 375,975 mm³/min.

EXAMPLE 7.7: CALCULATING THE CUTTING TIME IN FACE MILLING WHEN THE CUTTER IS OFFSET

By using the data in Example 7.6, calculate the cutting time for one milling pass if the cutter is offset to one side of the work so that the swath cut by the cutter is 30 mm wide.

Solution

$L = 320$ mm, $w_s = 30$ mm, $D = 40$ mm, $f_r = 557$ mm/min
    By using Equation 7.8,

$$L_A = L_o = \sqrt{w_s\,(D - w_s)} = \sqrt{30(40 - 30)} = 17.3\,\text{mm}$$

By using Equation 7.6,

$$T_c = \frac{L + L_o + L_A}{f_r} = \frac{320 + 17.3 + 17.3}{557} = 0.636\,\text{min} = 38.2\,\text{s}$$

The cutting time for one milling pass = 38.2 s.

EXAMPLE 7.8: COMPUTING METAL REMOVAL RATE IN FACE MILLING WHEN THE CUTTER IS OFFSET

By using the data in Example 7.7, compute the metal removal rate.

Solution

$d = 5$ mm, $w_s = 30$ mm, $f_r = 557$ mm/min
    By modifying Equation 7.3 for the special case of face milling with offset cutter position, we get:

$$\text{MRR} = w_s\,d\,f_r = 30 \times 5 \times 557 = 83{,}550\,\text{mm}^3/\text{min}$$

EXAMPLE 7.9: IDENTIFYING THE METHOD OF MILLING AND CALCULATING THE CUTTING TIME AND MRR WHEN THE CUTTING CONDITIONS ARE SHOWN IN A SKETCH/DRAWING

A peripheral milling operation is performed on the top surface of a rectangular 90-mm-wide work-part, as shown in Figure E-7.9. The cutting speed is 90 m/min. (a) Specify whether the method is up milling or down milling. (b) Calculate the time to make one pass across the surface. (c) Calculate the MRR.

Solution

a. The method is up milling because the cutter is rotating opposite to the work-piece motion.

b. $w = 90$ mm, $D = 65$ mm, $n_t = 22$, $L = 280$ mm, $d = 6$ mm, $f = 0.25$ mm/tooth, $v = 90$ m/min

$$A = \sqrt{d(D-d)} = \sqrt{6(65-6)} = 19.13\,\text{mm} \quad (\text{see Equation 7.5})$$

$$v = \pi D N$$

$$90,000 = \pi \times 65\,N$$

$$N = 440.68\,\text{rpm}$$

$$f_r = f \cdot N \cdot n_t = 0.25 \times 440.68 \times 22 = 2423\,\text{mm/min}$$

By using Equation 7.4,

$$T_c = \frac{L+A}{f_r} = \frac{280+19.13}{2423} = 0.123\,\text{min}$$

The time to make one pass = 0.123 min = 7.4 s.

c. MRR = $w\,d\,f_r$ = 90 × 6 × 2423 = 1,308,420 mm³/min.
The MRR = 1,308,420 mm³/min = 1308.42 cm³/min.

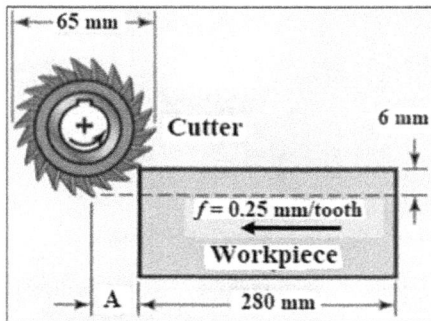

FIGURE E-7.9 Peripheral milling showing machining data

EXAMPLE 7.10: SELECTING THE APPROPRIATE SPECIFIC CUTTING FORCE BASED ON FEED/TOOTH

A medium-carbon steel 70-mm-wide workpiece is required to be milled at a cutting speed of 100 m/min with a depth of cut 2.2 mm using a table feed rate of 260 mm/min. The 200-mm outside-diameter cutter has 16 teeth. Select the appropriate specific cutting force for the milling operation (refer to Table 7.3).

Solution

$w = w_c = 70$ mm, $v = 100$ m/min, $d = 2.2$ mm, $f_r = 260$ mm/min, $D = 200$ mm, $n_t = 16$, $F_{sc} = ?$

Since the specific cutting force depends on the feed per tooth f_t, it is important to determine f_t.

By using Equation 7.1,

$$N = \frac{v}{\pi D} = \frac{100,000}{\pi \times 200} = 159.13 \, \text{rev/min}$$

By rewriting the modified form of Equation 7.2, we obtain:

$$f_t = \frac{f_r}{N n_t} = \frac{260}{159.13 \times 16} = 0.102 \, \text{mm/tooth}$$

By reference to Table 7.3, the feed = 0.1 mm/tooth corresponds to the specific cutting force of 1980 MPa for milling a medium-carbon steel work.

EXAMPLE 7.11: CALCULATING THE CUTTING POWER FOR MILLING OPERATION

By using the data in Example 7.10, calculate the cutting power for the milling operation, if the machine's efficiency is 85%.

Solution

$w = 70$ mm, $d = 2.2$ mm, $f_r = 260$ mm/min, $F_{sc} = 1980$ MPa, $\eta = 85\%$, $P_c = ?$

By using Equation 7.9,

$$P_c = \frac{d \cdot w \cdot f_r \cdot F_{sc}}{60 \times 10^6 \eta} = \frac{2.2 \times 70 \times 260 \times 1980}{60 \times 10^6 \times 0.85} = 1.55 \, \text{kW}$$

The cutting power = $P_c = 1.55$ kW.

EXAMPLE 7.12: CALCULATING THE TORQUE IN MILLING

By using the data in Example 7.11, calculate the torque in the milling operation.

Solution

$N = 159.13$ rev/min, $P_c = 1.55$ kW = 1550 W, $T = ?$

By using Equation 7.10,

$$T = \frac{P_c}{2\pi N} = \frac{1550}{2\pi \times 159.13} = 1.55\,\text{N}\cdot\text{m}$$

The torque in the milling operation = 1.55 N·m.

EXAMPLE 7.13: CALCULATING NUMBER OF REVOLUTIONS OF INDEX CRANK IN SIMPLE INDEXING

Calculate the number of rotations required for the index crank by simple indexing for the following divisions of work (blank): (a) 2 and (b) 5.

Solution
a. By using Equation 7.11,
Number of rotations of index crank required for each division of work
$= \frac{40}{Z} = \frac{40}{2} = 20$
b. Number of rotations of index crank required for each division of work
$= \frac{40}{Z} = \frac{40}{5} = 8$

EXAMPLE 7.14: IDENTIFYING THE PLATE'S COMPANY AND NUMBER IN SIMPLE INDEXING

Calculate the number of revolutions required for the index crank by simple indexing for the following divisions of work (blank): (a) 16 and (b) 29. Also identify the company and its index plate number.

Solution
a. By using Equation 7.11,
Number of turns of index crank required for each division of work
$= \frac{40}{Z} = \frac{40}{16} = 2\frac{8}{16}$
It means that for each division, the index crank must make 2 full revolutions and turn through 8 holes on the 16-holes circle of the Plate No. 1 of the Brown and Sharpe Company (see Table 7.4)
b. Number of turns of index crank required for each division of work
$= \frac{40}{Z} = \frac{40}{29} = 1\frac{11}{29}$
It means that for each division, the index crank must make 1 full revolutions and turn through 11 holes on the 29-holes circle of the Plate No. 2 patented by the Brown and Sharpe Company.

EXAMPLE 7.15: INDEXING FOR MILLING A GEAR WITH 48 TEETH – SIMPLE INDEXING

It is required to machine a gear with 48 teeth by using a milling machine. The available index plate has the following holes: 15, 16, 17, 18, 19, and 20 holes. Calculate the number of turns of the index crank using the simple indexing.

Solution

By using Equation 7.11,

Number of turns of index crank required for each division of work

$$= \frac{40}{Z} = \frac{40}{48} = \frac{5}{6} = \frac{5}{6} \times \frac{3}{3} = \frac{15}{18}$$

For each division, the index crank should turn through 15 holes on the 18-holes circle using the given index plate.

EXAMPLE 7.16: INDEXING FOR HEXAGONAL HEAD OF A BOLT WITH MULTIPLE OPTIONS

Calculate the number of revolutions of the index crank to machine a hexagonal head of a bolt using a milling machine. Identify at least three plates and the company.

Solution

Number of turns of index crank required for each division of work

$$= \frac{40}{Z} = \frac{40}{6} = 6\frac{2}{3}$$

The fraction can be engineered to obtain options for using three plates, as follows:

a. $\frac{2}{3} = \frac{2}{3} \times \frac{5}{5} = \frac{10}{15}$

It means that for each indexing, the index crank must make 6 full revolutions and turn through 10 holes on the 15-holes circle of the Plate No. 1 of the Brown and Sharpe Company.

b. $\frac{2}{3} = \frac{2}{3} \times \frac{7}{7} = \frac{14}{21}$

It means that for each indexing, the index crank must make 6 full revolutions and turn through 14 holes on the 21-holes circle of the Plate No. 2 of the Brown and Sharpe Company.

c. $\frac{2}{3} = \frac{2}{3} \times \frac{13}{13} = \frac{26}{39}$

It means that for each indexing, the index crank must make 6 full revolutions and turn through 26 holes on the 39-holes circle of the Plate No. 3 of the Brown and Sharpe Company.

EXAMPLE 7.17: INDEXING FOR GEAR WITH 55 TEETH – DIFFERENTIAL INDEXING

It is required to cut a gear with 55 teeth. (a) Determine the number of revolutions of the index crank, indicating the index plate # and the company, (b) select the differential change gears, and (c) indicate the direction of rotation of the index plate and the number of idle gears to be used.

The available change gears have the following number of teeth: 24,24,28,32,36,40,44,48,54,64.

Solution

Let us first try simple indexing using $Z = 55$

Number of revolutions of index crank required for each division $= \dfrac{Z_o}{Z} = \dfrac{40}{55}$

It is clear from the available index plates (Table 7.4) that 55 cannot be simple indexed.

a. $Z = 55$, $Z' = 54$ (see Rule 1 of differential indexing); $Z_o = 40$
By using Equation 7.12,

Number of revolutions of index crank required for each division $= \dfrac{Z_o}{Z'} = \dfrac{40}{54} = \dfrac{20}{27}$

It means that the index crank must turn through 20 holes on the 27-holes circle of the Plate No. 2 patented by the Brown and Sharpe Company (see Table 7.4).
b. For selecting the differential change gears, we use Equation 7.13,

$$\frac{Z_1}{Z_2} \cdot \frac{Z_3}{Z_4} = \frac{Z_o(Z'-Z)}{Z'} = \frac{40(54-55)}{54} = -\frac{40}{54} = -\frac{40}{54} \times \frac{24}{24}$$

Hence, the differential change gears are as follows: $Z_1 = 40$ teeth, $Z_2 = 54$ teeth, $Z_3 = 24$, and $Z_4 = 24$ teeth.
c. Since the change gear ratio is negative, the index plate rotation is in a direction opposite to that of the crank, and two idle gears must be used (see Rule 5 of differential indexing).

EXAMPLE 7.18: INDEXING FOR 67-TEETH GEAR – DIFFERENTIAL INDEXING

It is required to mill a gear with 67 teeth. Determine the number of revolutions of the index crank and select the differential change gears of the indexing head for cutting the gear.
The available index plate has the following holes: 15, 16, 17, 18, 19, and 20.
The available change gears are: 17, 20, 28, 36, 40, 48, 64, and 72.

Solution

$Z = 67$, $Z' = 68$, $Z_1 = ?$, $Z_2 = ?$, $Z_3 = ?$, $Z_4 = ?$

Number of revolutions of index crank required for each division $= \dfrac{Z_o}{Z'} = \dfrac{40}{68} = \dfrac{10}{17}$.

For each division, the index crank must turn through 10 holes on the 17-holes circle using the available index plate.
For determining the differential change gears, we use Equation 7.13,

$$\frac{Z_1}{Z_2} \cdot \frac{Z_3}{Z_4} = \frac{Z_o(Z'-Z)}{Z'} = \frac{40(68-67)}{68}$$
$$= \frac{40}{68} \times \frac{1}{1} = \frac{40}{4} \times \frac{1}{17} = \frac{40}{17} \times \frac{1}{4} = \frac{40}{17} \times \frac{18}{72} = \frac{20 \times 2}{17} \times \frac{18}{72}$$

$$\frac{Z_1}{Z_2} \cdot \frac{Z_3}{Z_4} = \frac{20}{17} \times \frac{36}{72}$$

Hence, the differential change gears are the following: $Z_1 = 20$ teeth, $Z_2 = 17$, $Z_3 = 36$, $Z_4 = 72$.

Since the change gears ratio is positive, the index plate rotation is in the same direction as that of the crank, and an idle gear is used.

EXAMPLE 7.19: INDEXING FOR MILLING A 119-TEETH GEAR – DIFFERENTIAL INDEXING

Calculate the number of revolutions of the index crank and select the differential change gears for cutting (milling) a gear with 119 teeth.

The available index plate has the following holes: 15, 16 17, 18, 19, and 20.

The available change gears are 24, 28, 28, 32, 36, 40, 44, 48, 56, 64, 72, and 100.

Solution

$Z = 119$, $Z' = 120$, $Z_1 = ?$, $Z_2 = ?$, $Z_3 = ?$, $Z_4 = ?$

Number of revolutions of index crank required for each division

$$= \frac{Z_o}{Z'} = \frac{40}{120} = \frac{1}{3} = \frac{1}{3} \times \frac{5}{5} = \frac{5}{15}.$$

For each division, the crank must turn through 5 holes on the 15-holes circle of the index plate.

$$\frac{Z_1}{Z_2} \cdot \frac{Z_3}{Z_4} = \frac{Z_o(Z'-Z)}{Z'} = \frac{40(120-119)}{120}$$

$$= \frac{1}{3} \times \frac{1}{1} = \frac{1}{3} \times \frac{28}{28} = \frac{24}{24 \times 3} \times \frac{28}{28} = \frac{24}{72} \times \frac{28}{28}$$

Hence, the differential change gears are $Z_1 = 20$ teeth, $Z_2 = 17$, $Z_3 = 36$, $Z_4 = 72$

The index plate rotation is in the same direction as that of the crank, and an idle gear is used.

EXAMPLE 7.20: INDEXING FOR MILLING A GEAR WITH 153 TEETH – DIFFERENTIAL INDEXING

It is required to mill a gear with 153 teeth. Determine the number of revolutions of the index crank and the differential change gears of the indexing head for cutting the gear.

The available index plate has the following holes: 15, 16, 17, 18, 19, and 20.

The available change gears are 28, 32, 36, 40, 48, 48, 54, and 64.

Solution

$Z = 153$, $Z' = 150$, $Z_1 = ?$, $Z_2 = ?$, $Z_3 = ?$, $Z_4 = ?$

Number of revolutions of index crank required for each division $= \dfrac{Z_o}{Z'} = \dfrac{40}{150} = \dfrac{4}{15}$

For each division, the crank must turn through 4 holes on the 15-holes circle of the index plate.

$$\frac{Z_1}{Z_2} \cdot \frac{Z_3}{Z_4} = \frac{Z_o(Z'-Z)}{Z'} = \frac{40(150-153)}{150} = -\frac{4}{15} \times \frac{3}{1} = -\frac{12}{15} = -\frac{2 \times 6}{3 \times 5} = -\frac{2}{3} \times \frac{6}{5}$$

$$\frac{Z_1}{Z_2} \cdot \frac{Z_3}{Z_4} = -\frac{2 \times 16}{3 \times 16} \times \frac{6 \times 8}{5 \times 8} = -\frac{32}{48} \times \frac{48}{40}$$

Hence, the differential change gears are $Z_1 = 32$ teeth, $Z_2 = 48$, $Z_3 = 48$, $Z_4 = 40$.

Since the change gear ratio is negative, the index plate rotation is in a direction opposite to that of the crank, and two idle gears must be used.

QUESTIONS AND PROBLEMS

7.1 Differentiate between the following terms with the aid of diagrams:
   a. peripheral milling and face milling and
   b. up milling and down milling.

7.2 List the milling operations that can be performed for each of the following types of milling machines:
   a. horizontal-spindle knee-and-column-type milling machine, and
   b. vertical-spindle knee-and-column-type milling machine.

7.3 Explain the following operations with the aid of sketches:
   a. end milling and
   b. side milling.

7.4 Draw a sketch showing a milling cutter's nomenclature.

7.5 How is direct indexing carried out in a milling machine practice? What is the limitation of the direct indexing?

P7.6 A tool steel work-part is to be machined by using a carbide milling cutter with 25-mm outside diameter.
   a. Select the cutting speed in mm/min and
   b. calculate the cutter rotational speed. Hint: Refer to Table 7.2.

P7.7 A brass work-part (200 mm × 60 mm × 15 mm) is to be machined by peripheral milling using a HSS cutter rotating at 120 rpm. The cutter has 20 teeth. It is required to remove 5 mm from the top surface of the work.
   a. Select the feed per tooth and
   b. calculate the MRR.

P7.8 By using the data in P7.7, calculate the cutting time for milling one pass by using the cutter with outside diameter of 28 mm.

P7.9 It is desired to remove 4 mm from the top surface of an aluminum rectangular workpiece (300 mm × 125 mm) in one pass by a face milling operation. The HSS cutter has 8 teeth and its outer diameter is 30 mm. The cutter is offset to one side of the work so that the swath cut by the cutter is 22 mm wide. Take the average value of the feed/tooth in Table 7.2. Calculate (a) the cutting time to make the single pass and (b) the metal removal rate.

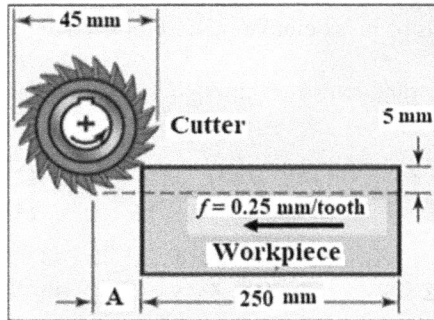

FIGURE P-7.11   The milling operation on the 80-mm wide work

P7.10 A gray cast iron 70-mm-wide workpiece is required to be milled at a cutting speed of 75 m/min with depth of cut 2 mm using a table feed rate of 200 mm/min. The 150-mm outside-diameter cutter has 12 teeth. Select the appropriate specific cutting force for the milling operation (refer to Table 7.3). Calculate the cutting power for the milling operation, if the machine's efficiency is 88%. Also compute the torque in the milling operation.

P7.11 A peripheral milling operation is performed on the top surface of a rectangular 80-mm-wide work-part, as shown in Figure P-7.11. The cutting speed is 75 m/min.
 a. Specify whether the method is up milling or down milling.
 b. Calculate the time to make one pass across the surface.
 c. Calculate the MRR.

P7.12 Calculate the number of revolutions required for the index crank by simple indexing for 18 divisions of work (blank). Also select the index plate and the company.

P7.13 Seven flutes are required to be cut in a cylindrical work. How many revolutions must the index crank make? Identify at least two plates and the company (see Table 7.4).

P7.14 a. Determine the number of revolutions of the index crank for cutting a gear with 53 teeth.
 b. Select the index plate (see Table 7.4), (c) Select the differential change gears.
 c. How many idle gears are needed? The available change gears are 24, 24, 28, 32, 36, 40, 44, 48, 56, 64, and 72.

P7.15 Determine the number of revolutions of the index crank for cutting a gear with 57 teeth.
 Select the index plate. Also select the differential change gears. How many idle gears are needed? The available change gears are 24, 29, 36, 40, 44, 48, 56, 64, and 72.

P7.16 Determine the number of revolutions of the index crank for cutting a gear with 127 teeth.

Select the index plate. Also select the differential change gears. How many idle gears are needed? The available change gears are 24, 28, 32, 36, 40, 44, 48, 56, and 64.

7.17 Underline the most appropriate answers for the following statements.

a. In which type of machine the spindle is mounted to a movable housing on the column?

(i) knee-and-column-type milling machine, (ii) universal horizontal milling machine, (iii) ram-type milling machine, (iv) none of them.

b. Which type of milling operation enables us machine the outside periphery of a flat part?

(i) profile milling, (ii) pocket milling, (iii) face milling (iv) slab milling.

c. In which type of milling operation, both sides of the work are machined (milled)?

(i) slot milling, (ii) slitting, (iii) end milling, (iv) straddle milling.

REFERENCES

Bray, S. (2004), *Milling*, Crowood Press Ltd., Wiltshire, UK.

Hall, H. (2013). *The Milling Machine for Home Machinists*, Fox Chapel Publishing, Mount Joy, PA.

Hayasaka, T., Ito, A., Shamoto, E. (2017), Generalized design method of highly-varied-helix end mills for suppression of regenerative chatter in peripheral milling. *Precision Engineering*, 48, 45–59.

Huda, Z. (2018), *Manufacturing: Mathematical Models, Problems, and Solutions*, CRC Press, Boca Raton, FL.

Vavruska, P., Zeman, P., Stejskal, M. (2018), Reducing machining time by pre-process control of spindle speed and feed-rate in milling strategies, *Procedia CIRP*, 77, 578–581.

Walker, J.R., Dixon, B. (2013), *Machining Fundamentals*, 9th Edition, Goodheart-Willcox, Tinley Park, IL.

8 Shaping/Planing Operations and Machines

8.1 SHAPING/PLANING OPERATION AND ITS APPLICATIONS

Shaping/planing is a machining operation that involves a linear relative motion between a workpiece and a single-point cutting tool to produce flat surface. In *shaping*, the tool has a reciprocating speed motion, whereas the workpiece has a feed motion; here the work is cross-fed during the machining operation (see Figure 8.1a). In *planing*, the workpiece has a reciprocating speed motion, whereas the tool has a feed motion; here the tool is cross-fed during the cut (see Figure 8.1b).

Shaping/planing operation can generate flat surfaces at different planes or orientations; these machining capabilities enable the operator to produce straight slots, grooves, pockets, T-slots, V-blocks, and the like (Colvin, 1943). Although milling operations are primarily carried out for generating flat surfaces by rotation of cutting tool, *shaping/planing* operation utilizes only reciprocating motion to provide the three necessary cutting conditions (cutting speed, feed, and depth of cut). The machine tool used to perform shaping operation is called a *shaper*, while the machine tool used for planing operation is known as a *planer*. *Shapers/planers* are one of the oldest machine tools used to generate flat surfaces by combination of relative linear motions of tool and workpiece (Burghardt, 2018). These machine tools are explained in the subsequent sections.

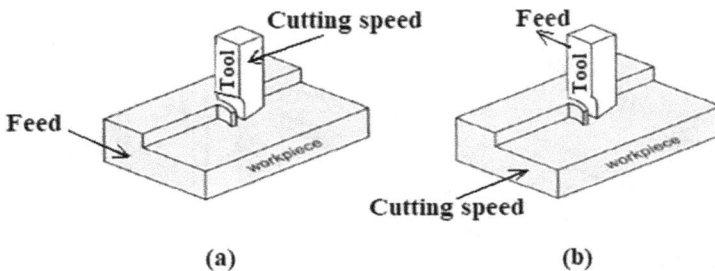

FIGURE 8.1   (a) Shaping operation and (b) planing operation

8.2 PLANING MACHINE TOOL – PLANER

8.2.1 PLANER – PARTS AND THEIR FUNCTIONS

A *planer* is a machine tool that produces planes and flat surfaces by reciprocating speed motion of the worktable, while the cutting tool has feed motion. A planer is similar to a shaper, but the former is larger and capable of doing heavy jobs. A planer consists of the following parts: bed, table, tool head, cross rail, column, and driving and feed mechanisms (see Figure 8.2). These parts are briefly discussed in the following paragraph.

A planer's *base* is a large and massive iron casting; it supports the column and all moving parts of the machine tool. The length of the base is almost double that of the table; this design feature permits free sliding movements of the table on the guideways of the base. The base contains the driving mechanism for the table. The *table* is a rectangular iron casting with smoothly machined top surface and T-slots to firmly and accurately hold the workpiece. The *column* is a rigid box-like structure vertically fastened on both sides of the base. The column houses the cross rail elevating screw as well as the vertical and cross-feed screws for tool heads. The cross rail is a strong casting that connects the two columns. The cross rail may be raised and lowered on the face of the column by means of elevating screw; it can be clamped at any position parallel to the top surface of the table. The *tool heads* are mounted on the cross rail by a saddle which provides cross feed motion to the tool head.

A planer can generate accurate flat surfaces in large workpieces; it can also cut slots (such as keyways). A planer can remove a tremendous amount of material in one pass with high accuracy *i.e.* material removal rate (MRR) in planing is higher than that in shaping (Parker, 2013).

8.2.2 WORKING PRINCIPLE OF PLANER

In order to operate a planer, the workpiece is firmly clamped to the table, and the single-point cutting tool is held in the tool head. Firstly, the tool head is kept

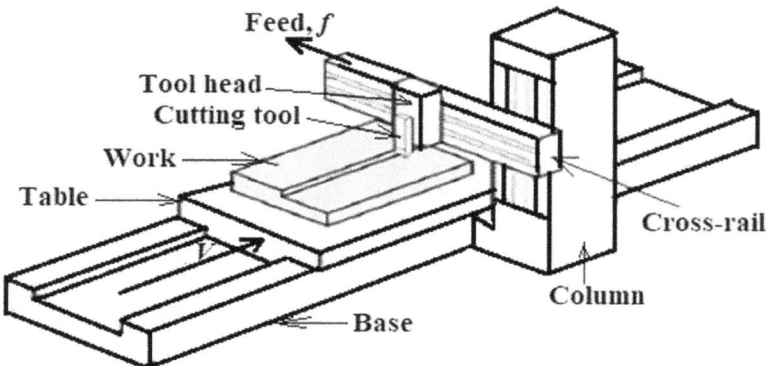

FIGURE 8.2   Schematic of a planer machine tool. (V = cutting speed, f = feed)

stationary, and the power is switched ON to give the reciprocating speed motion to the worktable. When the worktable moves forward, the tool cuts the material in the forward (cutting) stroke. When the table reaches the end of the forward stroke, it is slightly lowered so that the tool is cleared of the machined surface *i.e.* no material removal occurs in the return stroke. The reciprocating movements of the worktable continue unless power supply is switched OFF or the machine is interrupted.

8.3 SHAPING MACHINE TOOL – *SHAPER*

8.3.1 SHAPER – *PARTS AND THEIR FUNCTIONS*

The *shaper* is a machine tool that is used to produce flat surface by means of a straight line reciprocating single-point cutting tool. The flat surface so produced may be horizontal, vertical, or inclined at an angle. Based on the ram position and its travel, shapers may be horizontal, vertical, or travelling-head type shapers (Bradley, 1973). The *horizontal shaper* is the most commonly used shaping machine tool (see Figure 8.3). The principal components of a horizontal shaper are illustrated in the following paragraph.

The *base* is a heavy cast iron body that supports the *column* and the *bed*, which in turn support all the working parts, including the ram, the worktable, and the driving mechanism. The *column* is a cellular casting that houses driving motor and pulleys, variable speed gear box, the crank and slotted link mechanism, levers, handles, and the other controls of shaper. The ram slide ways are provided on the top of the *column*. The *ram* is a rigid casting that is located on the top of the column. The ram slides back and forth in square-shaped ways to transmit power to the cutting tool. The *ram* contains the stroke positioning mechanism and the down feed mechanism so as to enable the operator to adjust the starting point and the length of the stroke

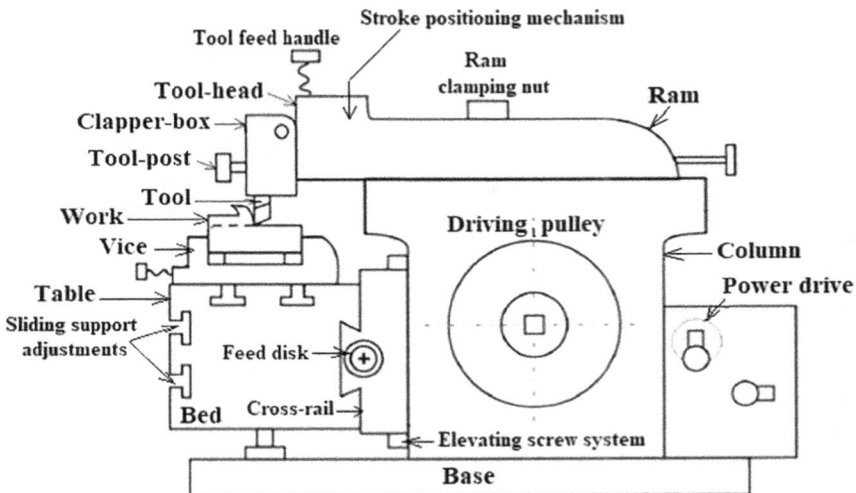

FIGURE 8.3 A schematic of horizontal shaper

(forward/reverse stroke). The *tool head* is fastened at the front of the ram; it holds the tool. The tool head can swivel from 0° to 90° in a vertical plane (see Figure 8.3). It is important to note that the cutting action is accomplished in the forward stroke of the ram, while the return stroke is idle. The *clapper-box* is an important component of a shaper because in its absence the cutting tool would drag over the work during the return stroke; this action would leave tool marks on the machined surface. This problem is overcome by use of a clapper-box. The clapper-box provides a lifting mechanism to clear the tool from the machined surface during the return stroke thereby avoiding damage to the machined surface on the return stroke. In modern shaping machines, the clapper-box is automatically raised by mechanical, pneumatic, or hydraulic action. The *cross rail* is a heavy casting attached to the front of the column on the vertical guideways. It carries the worktable slide ways. In order to compensate for different thickness of work, the cross rail can be raised or lowered by means of a screw lifting system. The *table* holds and supports the workpiece; it slides along the cross rail to provide feed to the work during shaping operation.

8.3.2 QUICK RETURN MECHANISM IN A SHAPER

We have learnt in the preceding subsection that, in a shaper, cutting occurs during the forward stroke of the ram, while the return stroke is idle. Hence, it is important for good productivity in shaper that the return stroke be faster than the cutting (forward) stroke. This objective is achieved by the *quick return mechanism* (*QRM*), which involves a crank and slotted bar link mechanism, as illustrated in Figure 8.4. In the *QRM*, the slotted bar is pivoted upon at its bottom end that is attached to the frame of column, while the upper end of the sliding bar is attached to the ram block by a pin. The slotted-bar sliding block is mounted upon the crankpin. The power is transmitted to the bull gear by a driving pinion, which receives its power from an electric motor. As the bull gear is driven, the crankpin revolves at a uniform speed. Thus, the sliding block fastened to the crankpin will slide up and down on the slotted bar (see Figure 8.4). As the sliding block will move inside the slotted bar, it provides a rocking movement, which is transferred to the ram that provides a reciprocating motion. Hence, the rotary motion of the bull gear or revolution of the crankpin is transmitted into the reciprocating motion of ram. In the *QRM*, the time taken on the return stroke is shorter than that of the forward (cutting) stroke; this quicker return stroke is justified by engineering analysis of the *QRM*, as illustrated in the following section (Section 8.4).

8.4 ENGINEERING ANALYSIS OF QRM IN A SHAPER

In the *QRM*, the fixed link *AB* forms the turning gear pair (driving pinion and bull gear), as shown in Figure 8.5. The driving crank *AC* revolves about the fixed center *A*. The sliding block (attached to the crankpin) at *C* slides along the slotted bar *BP* thereby causing BP to oscillate about the pivoted point B. The connecting rod *PQ* transmits the motion from *BP* to the ram (carrying the tool), which reciprocates along the length of stroke Q_1Q_2. The line of stroke Q_1Q_2 is perpendicular to the projected *BA*.

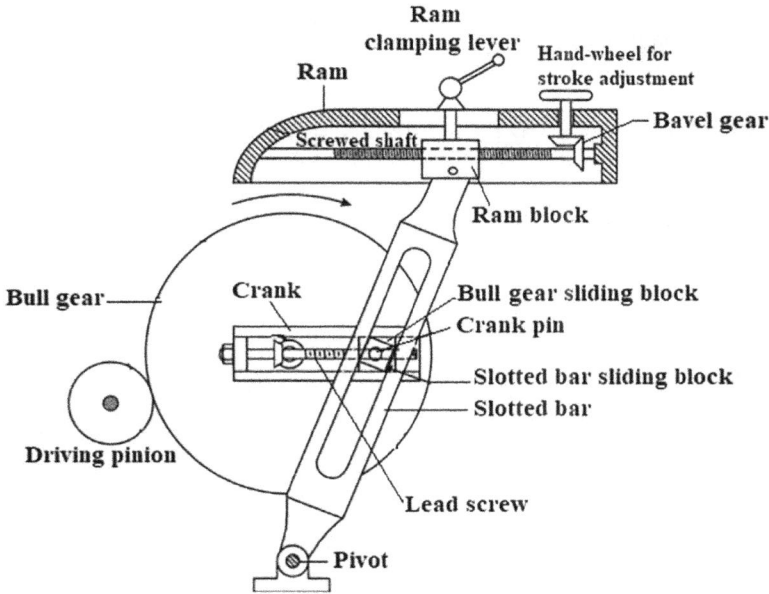

FIGURE 8.4 QRM in a horizontal shaper

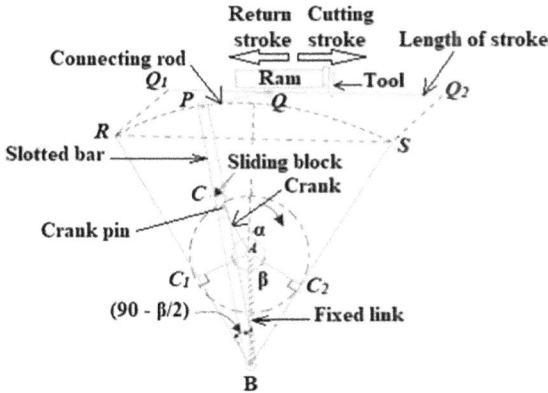

FIGURE 8.5 Engineering analysis of QRM in a crank shaper. (note: $\beta < \alpha$)

It is evident in Figure 8.5 that the cutting stroke occurs when the crankpin rotates from the position AC_1 to AC_2 (or through an angle α) in the clockwise direction. The return stroke occurs when the crankpin rotates from the position AC_2 to AC_1 (or through angle β) in the clockwise direction. Since $\beta < \alpha$, the return stroke is quicker than the cutting stroke; this is why the *QRM* is so named. Since the crank has uniform angular speed, the return/cutting ratio can be computed by:

$$\text{Return/cutting ratio} = k = \frac{\text{Return stroke time}}{\text{Cutting stroke time}} = \frac{\beta}{\alpha} = \frac{360° - \alpha}{\alpha} \qquad (8.1)$$

where α is the angle turned by the crankpin on the cutting stroke, and β is the angle turned by the crankpin on the return stroke (see Examples 8.1–8.2). Since $\beta < \alpha$, the ratio $k < 1$.

The maximum length of stroke $\overline{Q_1Q_2}$ can be calculated by referring to Figure 8.5 as follows:

$$\text{Maximum length of stroke} = L_{s(max)} = \overline{Q_1Q_2} = 2\,\overline{BP}\left(\frac{\overline{AC}}{\overline{AB}}\right) \qquad (8.2)$$

where $\overline{BP}$ is the length of the slotted bar, $\overline{AC}$ is the crank length, and $\overline{AB}$ is the length from the fixed (turning pair) center A to the pivoted end B of the slotted bar (see Example 8.3).

By using Figure 8.5, the angle turned by the crankpin on return stroke (β) can be computed by:

$$\sin\left(90 - \frac{\beta}{2}\right) = \frac{AC_1}{AB} = \frac{AC}{AB} \qquad (8.3)$$

The significance of Equation 8.3 is illustrated in Example 8.4.

8.5 ENGINEERING ANALYSIS OF SHAPING OPERATION

The starting and finishing positions, during the ram strokes, require some clearance/approach distance at each end of the workpiece, as shown in Figure 8.6. The stroke length (L_s) of the shaper should be adjusted based on the approach distance and the workpiece length as follows:

$$L_s = L + 2A \qquad (8.4)$$

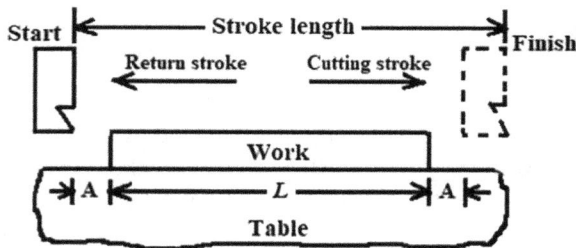

FIGURE 8.6 Stroke length, work length, and approach distances (A) on a horizontal shaper

where L is the length of the workpiece, and A is the approach distance (see Example 8.6).

The cutting speed on horizontal shapers is the average speed of the tool during the cutting stroke; it mainly depends on the number of ram (double) strokes per minutes (N) and the length of the stroke (L_s). The number of ram strokes per minute (N) is same as the rotational speed (rpm) of the bull gear (see Figure 8.4). The cutting speed is calculated by (Huda, 2018):

$$V_a = N L_s \left(1 + k\right) \qquad (8.5)$$

where V_a is the average cutting speed, mm/min; N is the number of double strokes per minute; L_s is the length of stroke, mm; and k is the return/cutting ratio (see Example 8.7).

The number of double strokes to complete the shaping job (n) can be computed by:

$$n = \frac{w}{f} \qquad (8.6)$$

where w is the width of the work, mm and f is feed, mm/double strokes (see Example 8.8).

The cutting time for a shaping job can be computed by:

$$T_c = \frac{n}{N} \qquad (8.7)$$

where T_c is the cutting time, min; n is the number of double strokes to complete the shaping job; and N is the number of double strokes per minute (see Example 8.9).

The MRR depends on the three cutting conditions: cutting speed V, feed f, and the depth of cut d, as shown in Figure 8.7.

The MRR is calculated by:

$$\text{MRR} = V_a \, d \, f \qquad (8.8)$$

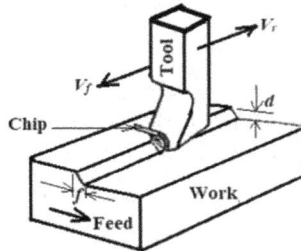

FIGURE 8.7 The three cutting conditions in shaping operation. (f = feed, d = depth of cut, V_f = cutting velocity on the forward stroke, V_r = cutting velocity on the return stroke)

MRR is the material removal rate, mm³/min; V_a is the cutting speed, mm/min; d is the depth of cut, mm; and f is the feed, mm/double strokes (see Example 8.10).

8.6 CALCULATIONS – *WORKED EXAMPLES ON SHAPING/SHAPER*

EXAMPLE 8.1: CALCULATING THE CUTTING RATIO IN SHAPING WITH QRM

It takes 10 s for the ram of a shaper to complete forward stroke, and 7 s on the return stroke. What is the return/cutting ratio?
Cutting stroke time = 10 s, Return stroke time = 7 s, k = ?
By using Equation 8.1,

$$k = \frac{\text{Return stroke time}}{\text{Cutting stroke time}} = \frac{7}{10} = 0.7$$

The return/cutting ratio = 0.7 or 7/10

EXAMPLE 8.2: COMPUTING THE ANGLE THE CRANKPIN TURNS ON THE CUTTING STROKE

By using the data in Example 8.1, calculate the angle through which the crankpin turns on the cutting stroke of the shaper.
$k = 0.7$, α = ?
By using Equation 8.1,

$$k = \frac{360° - \alpha}{\alpha}$$

$$0.7 = \frac{360° - \alpha}{\alpha}$$

$$0.7\alpha = 360 - \alpha$$

$$\alpha = 212°$$

The angle through which the crankpin turns on the cutting stroke = $\alpha = 212°$

EXAMPLE 8.3: COMPUTING THE MAXIMUM LENGTH OF STROKE WHEN CRANK LENGTH IS GIVEN

A shaper with a QRM has a 360-mm-long slotted bar and crank length of 75 mm. The length of the link from the fixed (turning pair) center to the pivoted end of the slotted bar is 200 mm. Calculate the maximum length of the stroke.

Solution

$$\overline{BP} = 360\,\text{mm}, \overline{AC} = 75\,\text{mm}, \overline{AB} = 200\,\text{mm}, \overline{Q_1Q_2} = ?$$

By using Equation 8.2,

$$L_{s(\max)} = \overline{Q_1 Q_2} = 2\,\overline{BP}\left(\frac{\overline{AC}}{\overline{AB}}\right) = \frac{2 \times 360 \times 75}{200} = 270\,\text{mm}$$

The maximum length of stroke = 270 mm = 27 cm.

EXAMPLE 8.4: COMPUTING THE ANGLE TURNED BY CRANKPIN ON RETURN STROKE

By using the data in Example 8.3, calculate the angle turned by the crankpin on return stroke.

Solution
By using Equation 8.3,

$$\sin\left(90 - \frac{\beta}{2}\right) = \frac{AC}{AB} = \frac{75}{200} = 0.375$$

$$\left(90 - \frac{\beta}{2}\right) = \sin^{-1} 0.375 = 22°$$

$$\frac{\beta}{2} = 90 - 22 = 68$$

$$\beta = 136°$$

The angle turned by the crankpin on return stroke = $\beta = 136°$

EXAMPLE 8.5: CALCULATING THE CUTTING RATIO WHEN THE TIMES DURATIONS ARE UNKNOWN

By using the data in Examples 8.3–8.4, calculate the cutting ratio for the shaping operation.

Solution
$\beta = 136°$, $k = ?$
  By reference to Figure 8.3, $\alpha + \beta = 360$

$$\alpha + 136 = 360$$

$$\alpha = 224°$$

By using Equation 8.1,

$$\text{Return/cutting ratio} = k = \frac{360° - \alpha}{\alpha} = \frac{360 - 224}{224} = 0.6$$

The return/cutting ratio = 0.6 = 6:10 = 3:5 or 3/5

EXAMPLE 8.6: CALCULATING THE STROKE LENGTH OF A SHAPER FOR A SHAPING OPERATION

The workpiece dimensions in a shaping operation are 270 mm × 180 mm × 90 mm. The approach distance at each end is 60 mm. The return/cutting ratio of the horizontal shaper is 3/5. The number of ram strokes is 35 double strokes/min, and the feed is 2.3 mm/double strokes. The depth of cut is 0.7 mm. What should be the stroke length of the shaper?

Solution
$L = 270$ mm, $A = 60$ mm, $L_s = ?$
  By using Equation 8.4,

$$L_s = L + 2A = 270 + (2 \times 60) = 390 \, \text{mm}$$

The stroke length of the shaper = 390 mm

EXAMPLE 8.7: CALCULATING THE CUTTING SPEED ON HORIZONTAL SHAPER

By using the data in Example 8.6, calculate the cutting speed of the tool.

Solution
$k = 3/5 = 0.6$, $L_s = 280$ mm, $N = 35$ double strokes/min, $V_a = ?$
  By using Equation 8.5,

$$V_a = N L_s (1 + k) = 35 \times 280 \times (1 + 0.6) = 15{,}680 \, \text{mm/min}$$

The tool's cutting speed = 15,680 mm/min = 15.68 m/min

EXAMPLE 8.8: CALCULATING THE NUMBER OF DOUBLE STROKES TO COMPLETE A SHAPING JOB

By using the data in Example 8.6, calculate the number of double strokes to complete the shaping job.

Solution
$w = 180$ mm, $f = 2.3$ mm/double strokes, $n = ?$
  By using Equation 8.6,

$$n = \frac{w}{f} = \frac{180}{2.3} = 78.3$$

The number of double strokes to complete the shaping job = 78.3

EXAMPLE 8.9: CALCULATING THE CUTTING TIME FOR A SHAPING JOB

By using the data in Examples 8.6–8.8, calculate the cutting time for the shaping job.

Solution

$n = 78.3$ double strokes, $N = 35$ double strokes/min, $T_c = ?$
By using Equation 8.7,

$$T_c = \frac{n}{N} = \frac{78.3\,\text{double strokes}}{35\,\text{double strokes/min}} = 2.24\,\text{min}$$

The cutting time for the shaping job = 2.24 min.

EXAMPLE 8.10: CALCULATING THE MRR IN THE SHAPING OPERATION

By using the data in Examples 8.6–8.7, calculate the MRR in the operation.

Solution

$V_a = 15,680$ mm/min, $d = 0.7$ mm, $f = 2.3$ mm/double strokes, MRR = ?
By using Equation 8.8,

$$\text{MRR} = V_a\, d\, f = 15,680 \times 0.7 \times 2.3 = 25,245\,\text{mm}^3/\text{min}$$

The MRR = 25,245 mm³/min = 25.24 cm³/min.

EXAMPLE 8.11: COMPUTING CUTTING SPEED, CUTTING TIME, AND MRR IS SHAPING OPERATION

A shaper working on a 45-cm stroke length makes 30 double strokes/ min with feed $f = 1.5$ mm/double strokes. The depth of cut is 0.5 mm, and return/cutting ratio is 3/4. This shaper is used to machine a 10-cm-wide metallic workpiece. Calculate the (a) cutting speed of tool, (b) metal removal rate, and (c) cutting time in the shaping operation.

Solution

$k = 0.75$, $L_s = 450$ mm, $N = 30$ double strokes/min, $d = 0.5$ mm, $w = 100$ mm, $V_a = ?$, MRR = ?, $T_c = ?$
a. By using Equation 8.5,

$$\text{Cutting speed} = V_a = N\, L_s \left(1 + k\right) = 30 \times 450 \times \left(1 + 0.75\right) = 23,625\,\text{mm/min}$$

b. By using Equation 8.8,

$$\text{Metal removal rate} = \text{MRR} = V_a\, d\, f = 23,625 \times 0.5 \times 1.5 = 17,719\,\text{mm}^3/\text{min}$$

c. By using Equation 8.6,

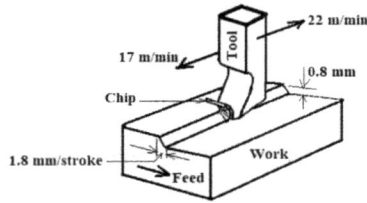

FIGURE E-8.12   Cutting conditions data in the shaping operation

$$n = \frac{w}{f} = \frac{100}{1.5} = 66.67 \, \text{double strokes}$$

By using Equation 8.7,

$$\text{Cutting time} = T_c = \frac{n}{N} = \frac{66.67 \, \text{double strokes}}{30 \, \text{double strokes/min}} = 2.22 \, \text{min}$$

EXAMPLE 8.12: CALCULATING THE MRR AND No. OF DOUBLE STROKES USING DATA FROM SKETCH

Refer to Figure E-8.12. Calculate the MRR and the number of double strokes to complete the shaping job on a workpiece with dimensions 280 × 170 × 70 mm³.

Solution
V_f = 17,000 mm/min, V_r = 22,000 mm/min, d = 0.8 mm, f = 1.8 mm/double strokes, w = 170 mm

$$V_a = \frac{V_f + V_r}{2} = \frac{17,000 + 22,000}{2} = 19,500 \, \text{mm/min}$$

By using Equation 8.8,

Metal removal rate = $MRR = V_a \, d \, f = 19,500 \times 0.8 \times 1.8 = 28,080 \, \text{mm}^3/\text{min}$.

By using Equation 8.6,

Number of double strokes to complete the shaping job = $n = \frac{w}{f} = \frac{170}{1.8} = 94.5$

QUESTIONS AND PROBLEMS

8.1 Differentiate between shaping and planing machining operations with the aid of diagrams.

8.2 Draw a labeled sketch of a horizontal shaper.

8.3 Refer to a shaper's *QRM*. Why does the sliding block (fastened to the crank-pin) slide up and down on the slotted bar when the driving pinion receives power from a motor? Explain.

8.4 Draw a diagram of a shaper's *QRM* showing a smaller angle turned by the crankpin on the return stroke as compared to the angle turned during the cutting stroke.

8.5 List the main parts of a planer. Briefly explain the working principle of a planer.

P8.6 In a shaper's QRM, the angle turned by the crankpin on return stroke is 140°. Calculate the return/cutting ratio for the shaping operation.

P8.7 A shaper with a QRM has a 300-mm-long slotted bar and a crank length of 66 mm. The length of the link from the fixed (turning pair) center to the pivoted end of the slotted bar is 170 mm. Calculate the maximum length of the stroke.

P8.8 The workpiece dimensions in a shaping operation are 320 mm × 200 mm × 100 mm. The approach distance at each end is 65 mm. The return/cutting ratio of the horizontal shaper is 3/4. The number of ram strokes is 40 double strokes/min., and the feed is 1.8 mm/double strokes. The depth of cut is 0.8 mm. Calculate the (a) cutting speed of tool, (b) metal removal rate, and (c) cutting time in the shaping operation.

P8.9 It takes 12 s for the ram of a shaper to complete forward stroke, and 8 s on the return stroke. Calculate the (a) return/cutting ratio and (b) angle through which the crankpin turns on the cutting stroke of the shaper.

P8.10 By using the data in P8.8, draw a sketch of the shaping operation showing the data for the three cutting conditions: the cutting speed, the feed, and the depth of cut.

REFERENCES

Bradley, I. (1973), *The Shaping Machine*, Model and Allied Publications, Madison, WI, USA.

Burghardt, H.D. (2018), *Machine Tool Operation*, Vol. 2 : Drilling Machine, Shaper and Planer, Milling and Grinding Machines, Spur Gears and Bevel Gears, Forgotten Books, London.

Colvin, F.H. (1943), *Planing, Shaping, and Slotting*, McGraw Hill Company Inc., New York.

Huda, Z. (2018), *Manufacturing – Mathematical Models, Problems, and Solutions*, CRC Press, Boca Raton, FL.

Parker, D.T. (2013), *Building Victory: Aircraft Manufacturing in the Los Angeles Area in World War II*, 201, 73, Cypress, CA, USA.

9 Broaching and Broach Design

9.1 BROACHING AND ITS APPLICATIONS

9.1.1 BROACHING – AN INTRODUCTION

Broaching is a machining operation that involves the removal of metal by use of a multipoint cutting tool with teeth arranged in a row. The broaching machine tool is called a *broaching machine*, and the broaching cutting tool is called a *broach*. In broaching, the broach moves linearly relative to workpiece in the direction of broach axis. The teeth in the broach are so arranged that each tooth is successively higher than the previous tooth. Broaching enables us to broach internal or external surfaces. Figure 9.1a illustrates broaching for internal work shape; here the internal (round) broach is inserted into a pilot or starting hole which is made on the workpiece before machining. The broaching operation for external surfaces is shown in Figure 9.1b. A simple manual broaching job, such as cutting a keyway in a part, requires only a broach, an arbor press, and the appropriate fixtures. However, production broaching requires specialized broaching machines and is suitable for a large number of parts.

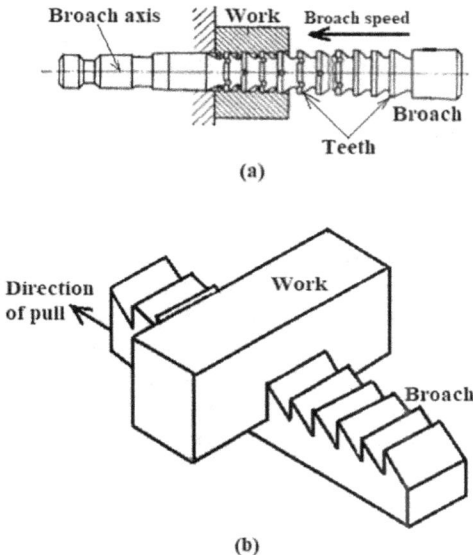

FIGURE 9.1 A schematic of broaching operation: (a) internal broaching and (b) external broaching

9.1.2 APPLICATIONS OF BROACHING

Broaching operation can cut almost any work shape from simple flats and slots to gears and turbine blade hubs for aircraft engines. Broaching is a highly productive method of machining a workpiece requiring precision and accuracy; it is applicable to many materials, including ferrous and non-ferrous metals and even some plastics. Broaching machines can produce both large and small parts, such as hand tools, appliances, automotive, farm implements, turbines, plumbing, military applications, and the like. There are many advantages of the broaching operation; these include close tolerance, good surface finish, variety of external and internal work shapes, rough to finish machining in one pass, less cutting time, and the like. However, the tooling is expensive owing to the complicated and custom shaped geometry of the broach. In order to justify the high tooling cost, the broaching process is suitable only for large volume of production. Additionally, the work material must be strong enough to withstand high cutting forces in broaching.

9.2 BROACH AND ITS DESIGN ANALYSIS

9.2.1 BROACH AND ITS PARTS

The broach is a multipoint cutting tool with a progressive and linear cutting edges. A typical broach consists of many rows of teeth (at the begin end) that do roughing, then a few rows of teeth (in the middle) for semi-finishing, and another few rows (at the finish end) that finish-machine the surface (Sharma, 2008). This is why the broaching operations ever finish the surface in a single pass. The important parts of a broach are shown in Figure 9.2.

9.2.2 BROACH MATERIALS AND TYPES

Broach Materials. The broach design is based on the workpiece and broach geometries, the properties of the broach and workpiece materials, and other related factors.

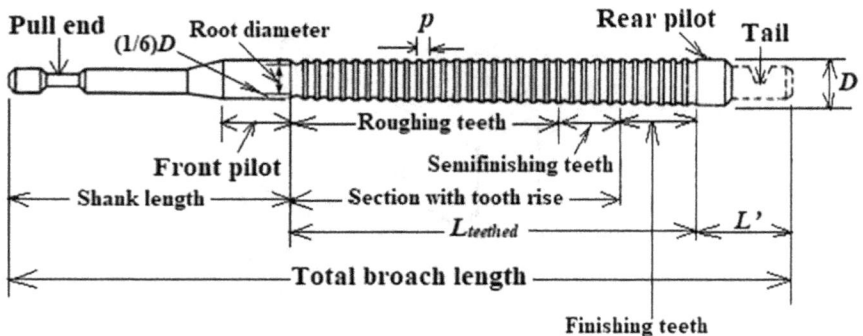

FIGURE 9.2 A broach with its parts showing geometric design parameters. (p = pitch, D = broach diameter)

Broaches are generally made of high speed steel (HSS), but superior quality broaches are made of cemented carbide. For advanced applications, broaches are coated with titanium nitride or cubic boron nitride (CBN) to extend their tool lives and to increase superficial hardness. The shape of the machined surface is determined by the contour of the cutting edges on the broach. For example, cylindrical/round broaches are used for internal cutting by compression, whereas plain broaches are employed for external cutting by traction. However, broaches can also be built and categorized by other means.

Types of Broach. Depending on the applied tensile/compressive force, a broach may be either a *push broach* or a *pull broach*. A *pull broach*, during broaching operation, is subjected to tensile force, which helps in maintaining alignment and prevents buckling. Pull broaches are generally made as a long single piece; they are particularly used for internal broaching. A *push broach* is a column in compression. Push broaches are relatively short in length (to avoid buckling) and may be made in segments. Push broaches are generally used for external broaching.

9.2.3 Design Analysis of Broach

A *push broach* must be relatively short since it will buckle under excessive forces. The maximum force that can be applied on a *push broach* can be computed by (Linsley, 1961):

$$F_{max} = \frac{S_y D_r^4}{L^2} \tag{9.1}$$

where F_{max} is the maximum force, N; S_y is the yield strength of the broach material, MPa; D_r is the minimum root diameter, mm; and L is the length from push end to the first tooth, mm (see Example 9.1).

The geometric design of a broach is based on mathematical relationships between pitch, broach length, broach diameter, and the like. The pitch of an internal broach can be calculated by:

$$p = k \sqrt{l_c} \tag{9.2}$$

where p is the broach pitch, mm, l_c is the length of cut per tooth, mm/tooth, and $k = 1.25 - 1.5$.

The length of the teethed part ($L_{teethed}$) can be calculated by:

$$L_{teethed} = p\left(n_r + n_s + n_f\right) \tag{9.3}$$

where P is the pitch, and n_r, n_s, and n_f are the number of roughing teeth, semi-finishing teeth, and finishing teeth, respectively (see Example 9.2).

The cutting length is the length of the section with tool rise (see Figure 9.2). The cutting length (L_c) is related to the length of the teethed part and the length of the finishing teeth, as follows:

$$L_c = L_{teethed} - L_f \tag{9.4}$$

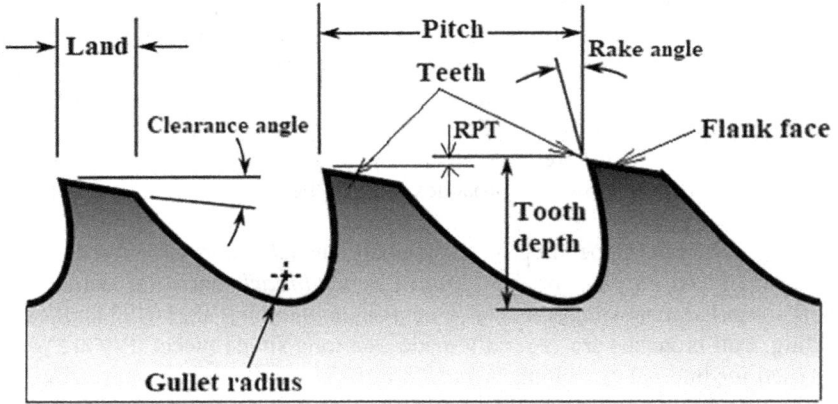

FIGURE 9.3 Terminologies/design parameters for teeth in a broach. (RPT = rise per tooth)

where L_f is the length section comprising of finishing teeth (see Example 9.3).

It is evident in Figure 9.2 that the root diameter (D_r) is related to the broach diameter (D) by:

$$D_r + 2\left(\frac{1}{6}D\right) = D \tag{9.5}$$

The rearrangement of terms in Equation 9.5 yields the expression for the root diameter as follows:

$$D_r = \frac{2}{3}D \tag{9.6}$$

where D_r is the root diameter and D is the broach diameter (see Example 9.4).

The geometric design parameters for teeth in a broach include the pitch, the tooth depth, the land, the gullet radius, and the like, as shown in Figure 9.3. The tooth depth, the land, and the gullet radius are related to the pitch by the following expressions:

$$\text{Tooth depth} = 0.4\, p \tag{9.7}$$

$$\text{Land} = \text{gullet radius} = 0.25\, p \tag{9.8}$$

The significance of Equations 9.7–9.8 is illustrated in Example 9.5.

9.3 ENGINEERING ANALYSIS OF BROACHING OPERATION

The cutting speeds in industrial broaching process, by using a HSS broach, range from 2 to 20 m/min (Schulze et al., 2012). The cross-sectional area of uncut chip per tooth (CPT) depends on the type of broach. The area of uncut CPT for a keyway broach is calculated by:

$$A_c = f_z\, w \tag{9.9}$$

where A_c is the cross-sectional area of uncut CPT, mm²; f_z is the cut per tooth or RPT, mm (see Figure 9.3); and w is the keyway width of the broach, mm.

The area of uncut CPT for a round broach is given by:

$$A_c = f_z\left(\pi D\right) = \pi D f_z \tag{9.10}$$

where D is the diameter of round broach, mm. The RPT ($=f_z$) for round broach varies from 0.03 to 0.06 mm. Once the cutting speed V and the area of uncut CPT A_c are known, the metal removal rate per tooth can be calculated by:

$$MRR\,/\,\text{tooth} = A_c\,V \tag{9.11}$$

where MRR is the metal removal rate, mm³/min and V is the cutting speed, mm/min (see Example 9.6). The material removal rate (MRR) in one pass can be determined by:

$$MRR/\text{pass} = \left(MRR/\text{tooth}\right)\cdot z \tag{9.12}$$

where z is the number of teeth simultaneously in contact with the chip during broaching operation (see Example 9.7).

The CPT for round broaches is related to the cut per tooth or RPT by:

$$CPT = \frac{1}{2}\left(RPT\right) \tag{9.13}$$

The required force for a broaching operation with roughing teeth of a round broach can be calculated by (Rajam, 1997):

$$F_{br} = \pi\, z\, D_p\left(CPT_R\right)C \tag{9.14}$$

where F_{br} is the force required for broaching, lbf; z is the number of teeth simultaneously in contact with the chip during broaching; D_p is the diameter of pilot or starting hole, in; CPT_R is the CPT roughing, in; and C is the constant for rough broaching, $C = 99470$ (see Example 9.8).

9.4 CALCULATIONS – WORKED EXAMPLES ON BROACHING

EXAMPLE 9.1: COMPUTING THE MAXIMUM FORCE THAT CAN BE APPLIED ON A BROACH

A push broach has the minimum root diameter of 2 in, and the length from its push end to the first row is 8 in. The yield strength of the broach material is 90,000 psi. Calculate the maximum force (in Newton) that can be applied on the broach for a broaching operation.

Solution

$S_y = 90,000$ psi, $D_r = 2$ in, $L = 7.8$ in, $F_{max} = ?$

By using Equation 9.1

$$F_{max} = \frac{S_y D_r^4}{L^2} = \frac{90000 \times 2^4}{7.8^2} = \frac{90000 \times 16}{60.84} = 23,668.64 \text{ lbf}$$

$F_{max} = 23,668.64 \text{ lbf} = 4.448 \times 23,668.64 = 105,278 \text{ N} = 105.278 \text{ kN}$

EXAMPLE 9.2: COMPUTING THE PITCH AND THE LENGTH OF TEETHED SECTION OF A BROACH

The length of cut per tooth in a broaching operation is 1 cm. The number of teeth in the roughing section, the semi-finishing section, and the finishing section of the internal broach are 24, 7, and 6, respectively. Calculate the teethed length (in mm) of the broach.

Solution

$l_c = 1$ cm $= 10$ mm/tooth; $n_r = 24$, $n_s = 7$, $n_f = 6$, $k = 1.3$, $L_{teethed} = ?$

By using Equation 9.2,

$$p = k\sqrt{l_c} = 1.3\sqrt{10} = 4.111 \text{ mm}$$

By using Equation 9.3,

$$L_{teethed} = p(n_r + n_s + n_f) = 4.111(24 + 7 + 6) = 152.11 \text{ mm}$$

The teethed length of the broach $= L_{teethed} = 152.11$ mm.

EXAMPLE 9.3: CALCULATING THE CUTTING LENGTH OF A BROACH

By using the data in Example 9.2, calculate the cutting length of the broach.

Solution

The length of the finishing-teeth section $= L_f = (p)(n_f) = 4.11 \times 6 = 24.66$ mm

By using Equation 9.4,

$$L_c = L_{teethed} - L_f = 152.11 - 24.66 = 127.45 \text{ mm}$$

The cutting length of the broach $= 127.45$ mm

EXAMPLE 9.4: CALCULATING THE ROOT DIAMETER OF A BROACH

The diameter of an internal broach is 20 mm. What is the root diameter of the broach?

Solution
By using Equation 9.6,

$$D_r = \frac{2}{3}D = \frac{2}{3}(20) = 13.33\,\text{mm}$$

The root diameter of the broach = 13.33 mm

EXAMPLE 9.5: CALCULATING THE TOOL DEPTH, LAND, AND GULLET RADIUS OF A BROACH

The pitch of a surface broach is 3.5 mm. Calculate the following design parameters of the broach: (a) tooth depth, (b) land, and (c) gullet radius.

Solution
  a. By using Equation 9.7,

$$\text{Tooth depth} = 0.4\,p = 0.4 \times 3.5 = 1.4\,\text{mm}$$

  b. By using Equation 9.8,

$$\text{Land} = 0.25\,p = 0.25 \times 3.5 = 0.87\,\text{mm}$$

  c. Gullet radius = $0.25\,p = 0.25 \times 3.5 = 0.87$ mm

EXAMPLE 9.6: CALCULATING THE MRR PER TOOTH IN BROACHING

The cutting speed in a broaching operation is 10 m/min. The broach has keyway width of 8 mm and the RPT of 0.04 mm. Calculate the MRR per tooth.

Solution
V = 10 m/min = 10,000 mm/min, w = 8 mm, RPT = f_z = 0.04 mm, MRR/tooth = ?
  By using Equation 9.9,

$$A_c = f_z\,w = 0.04 \times 8 = 0.32\,\text{mm}^2$$

By using Equation 9.11,

$$\text{MRR/tooth} = A_c\,V = 0.32 \times 10000 = 3,200\,\text{mm}^3/\text{min}$$

EXAMPLE 9.7: CALCULATING THE MRR PER PASS IN BROACHING

By using the data in Example 9.6, calculate the MRR per pass in the broaching operation if four teeth are simultaneously in contact with the chip during broaching.

Solution

MRR/tooth = 3,200 mm³/min, $z = 4$, MRR/pass = ?

By using Equation 9.12,

$$\text{MRR/pass} = \left(\text{MRR/tooth}\right)z = 3,200 \times 4 = 12,800\,\text{mm}^3/\text{min}$$

The MRR per pass = 12,800 mm³/min = 12.8 cm³/min.

EXAMPLE 9.8: CALCULATING THE REQUIRED FORCE FOR A BROACHING OPERATION

The diameter of a pilot hole for a broaching operation is 0.8 in. The number of teeth of the round broach simultaneously in contact with the chip during broaching are 3. The RPT in the roughing section of the broach is 0.002 in. Calculate the required force for the broaching operation (in kilo Newton).

Solution

$D_p = 0.8$ in., $RPT_R = 0.002$ in, $z = 3$, $C = 99470$, $F_{br} = ?$

By using Equation 9.13,

$$CPT_R = \frac{1}{2}\left(RPT_R\right) = 0.5 \times 0.002 = 0.001\,\text{in}$$

By using Equation 9.14,

$$F_{br} = \pi\, z\, D_p \left(CPT_R\right) C = 3.142 \times 3 \times 0.8 \times 0.001 \times 99470 = 750\,\text{lbf}$$

The force required for the broaching operation = 750 lbf = 4.448 × 750 = 3,336.37 N = 3.34 kN

QUESTIONS AND PROBLEMS

9.1 Differentiate between internal broaching and external broaching, with the aid of diagrams.

9.2 a. Highlight the advantages and applications of broaching over other machining operations.

 b. Why is broaching generally practiced for mass production?

9.3 a. Draw a sketch of a broach showing its important parts/design parameters.

 b. Draw a sketch of a broach showing terminologies/design parameters for teeth.

 c. Distinguish between push broach and pull broach.

P9.4 The length of cut per tooth in a broaching operation is 8 mm. The number of teeth in the roughing section, the semi-finishing section, and the finishing section of the internal broach are 26, 8, and 6, respectively. Calculate the teethed length of the broach.

P9.5 A push broach has the minimum root diameter of 1.8 in, and the length from its push end to the first row is 7.5 in. The yield strength of the broach material is

92,000 psi. Calculate the maximum force (in Newton) that can be applied on the broach for a broaching operation.

P9.6 The cutting speed in a broaching operation is 10 m/min. The broach has keyway width of 8 mm and the RPT of 0.04 mm. Calculate the (a) MRR per tooth and (b) MRR per pass.

P9.7 The diameter of an internal broach is 18 mm. What is the root diameter of the broach?

P9.8 The diameter of a pilot hole for a broaching operation is 0.7 in. The number of teeth of the round broach simultaneously in contact with the chip during broaching is 3. The RPT in the roughing section of the broach is 0.0018 in. Calculate the required force for the broaching operation (in kilo Newton).

REFERENCES

Linsley, H.E. (1961), *Broaching Tooling and Practice*, Industrial Press Inc., New York.

Rajam, S.K. (1997), *Design and Finite Element Analysis of the Broaching Tools*, M.S Thesis, Ohio University's College of Engineering and Technology, Ohio.

Schulze, V., Meier, H., Strauss, T., Gibmeier, J. (2012), High speed broaching of case hardening steel SAE5120, *Procedia CIRP*, **1**: 431–436.

Sharma, P.C. (2008), *A Textbook of Manufacturing Technology-II*, S. Chand Publishing, New Delhi, India.

10 Gear Cutting/ Manufacturing

10.1 GEARS – *SPUR GEAR NOMENCLATURE*

Gears. Gears are crucial parts of many machines. A gear is a machine element that has teeth cut around cylindrical or cone-shaped surfaces with equal spacing. By meshing a pair of gears, rotational speed and power can be transmitted from the driving shaft to the driven shaft. Hence, gears help increase torque output. Gears' teeth are generally surface hardened to ensure a high wear resistance combined with toughness (Huda, 2020). An important design parameter of a gear is its *module*. A gear's module is defined as the pitch diameter divided by the number of teeth. A variety of gear types are used as machine elements in motors and machineries. Some basic types of gears include spur gear, helical gear, bevel gears, worm gears, and the like (Lelikov, 2009). The teeth on spur gears are straight whereas the teeth on helical gears are angled with a helix angle. Since the scope of this book does not permit the explanation of all the types of gears, only spur gears are briefly described in the following paragraph.

Spur Gears. Spur gears are the most commonly used gears; they are used in series for large gear reductions. The teeth on spur gears are straight running parallel to the gear axis (see Figure 10.1a). Spur gears are used in washing machines, windup alarm clocks, and other similar devices. Spur gears have a normal gear ratio in the range of

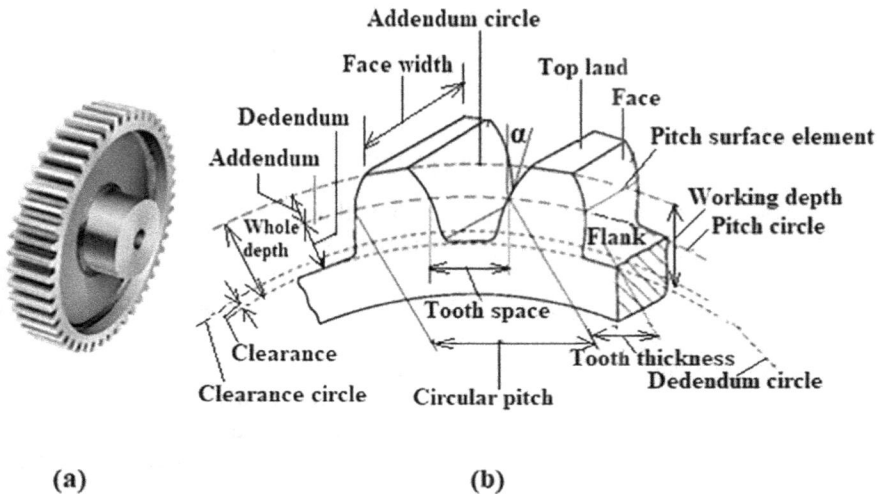

FIGURE 10.1 Spur gear (a) and nomenclature of a spur gear (b). (α = pressure angle)

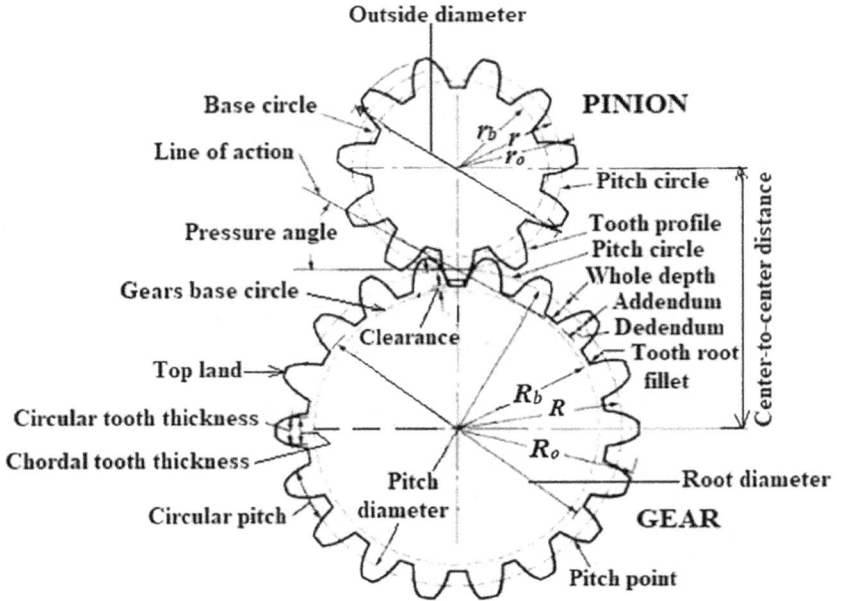

FIGURE 10.2 Nomenclature of a pinion-gear pair. (R_o = gear outside radius, R_b = gear base radius, R = gear pitch radius, r_o = pinion outside radius, r_b = pinion base radius, r = pinion pitch radius)

1:1–6:1. The nomenclature of a spur gear is illustrated in Figure 10.1b. The distance from the pitch circle to the addendum circle is equal to the addendum or *module* of spur gear.

Since a gear is meshed with a pinion in industrial practice, it is useful to consider the nomenclature of a pair of gears, as illustrated in Figure 10.2.

10.2 GEAR CUTTING METHODS/PROCESSES

Gears can be cut/manufactured by a number of methods; all the methods of gear cutting may be broadly divided into the following two principal groups: (1) gear forming and (2) gear generation. *Gear forming* involves machining by use of a form tool i.e. the machined gear's geometry is determined by the shape and geometry of the cutting tool. There are two commonly used gear forming operations: (a) gear milling and (b) gear broaching. Gear generation involves machining that results in the machined part's geometry dictated by the relative motion of workpiece and tool. Two generally used gear generation operations include (i) gear shaping and (ii) gear hobbing. Figure 10.3 shows the classification chart of gear cutting processes, which are explained in the subsequent sections.

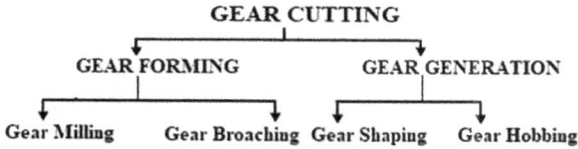

FIGURE 10.3 Classification chart of gear cutting processes

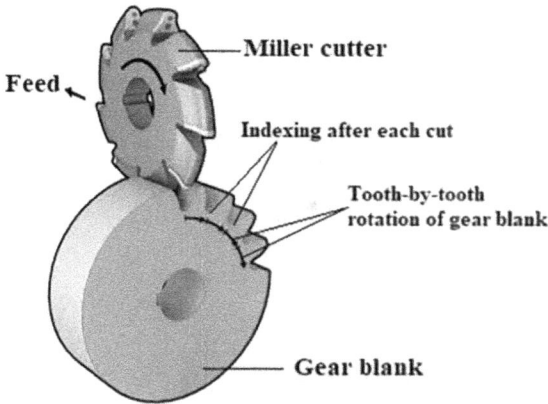

FIGURE 10.4 Gear milling process

10.3 GEAR MILLING

10.3.1 GEAR MILLING PROCESS

Gear milling involves the use of a form milling cutter to produce teeth in a work-piece – *gear blank*. In gear milling, a rotating cutter feeds axially along the face width of the gear blank at appropriate depth to produce a gear tooth. Once a tooth has been cut, the cutter is withdrawn and the gear blank is rotated (indexed). Then the cutter performs another tooth-cutting operation (see Figure 10.4). The process continues until all teeth on the gear blank are cut. In case of helical gear cutting, the cutter is angled. Figure 10.4 shows that gear milling involves indexing after each cut; the indexing in milling and its calculations have been explained in Chapter 7.

10.3.2 ENGINEERING ANALYSIS OF GEAR MILLING

Gear cutting/manufacturing, in industrial practice, requires effective application of gear milling cutters for a stable machining process. This requirement demands the determination of cutting power and torque, which in turn require computation for cross-sectional chip area, the specific cutting force, and the tooth space area. Since the arc of engagement of a gear milling cutter is generally very short, the portion of the cutter engaged in the gear blank is very small (see Figure 10.4). By knowing the cutting speed and the milling cutter's diameter, the rotational speed of the cutter (N) can be computed by using Equation 7.1. By knowing the feed per tooth (f_t) and the number of teeth on the cutter (n_t), the feed rate (f_r) can be calculated by using Equation 7.2. The average chip thickness is calculated by (Marsh, 2016):

$$h_m = \frac{2\,a_e\,f_t}{D_c \cos^{-1}\left(1 - \dfrac{2\,a_e}{D_c}\right)} \tag{10.1}$$

where h_m is the average chip thickness, mm; a_e is the working engagement, mm; and D_c is the cutter's outside diameter, mm (see Example 10.1). For a one pass operation for spur gear cutting, $a_e = H =$ whole depth of tooth space (see Figures 10.1–10.2).

Once h_m has been computed by using Equation 10.1, the specific cutting force (with m_c factored in) can be calculated by:

$$k_c = \frac{k_{c1}}{h_m^{\,m_c}} \tag{10.2}$$

where k_c is the specific cutting force with m_c factored in; k_{c1} is the specific cutting force for a given material, N/mm²; and m_c is the factor of specific rise in cutting force relative to chip thickness (see Example 10.2). The k_{c1} data for milling with small engagements for various materials is presented in Table 10.1.

TABLE 10.1

Specific Cutting Force (k_{c1}) Data for Milling with Small Engagements (Coromant, 2010)

ISO P MC No.	CMC No.	Material	k_{c1} (N/mm²)	Brinell Hardness BHN	m_c
		Un-alloyed Steel			
P1.1.Z.AN	01.1	0.10–0.25%C	1500	125	0.25
P1.2.Z.AN	01.2	0.25–0.55%C	1600	150	0.25
P1.3.Z.AN	01.3	0.55–0.8%C	1700	170	0.25
P1.3.Z.AN	01.4		1800	210	0.25
P1.3.Z.AN	01.5		2000	300	0.25
		Low-alloy steel (alloying elements ≤ 5%)			
P2.1.Z.AN	02.1	Non-hardened	1700	175	0.25
P2.5.Z.HT	02.2	Hardened and tempered	1900	300	0.25
		High-alloy steel (alloying elements > 5%)			
P3.0.Z.AN	03.11	Annealed	1950	200	0.25
P3.1.Z.AN	03.13	Hardened tool steel	2150	200	0.25
P3.0.Z.HT	03.21		2900	300	0.25
P3.1.Z.HT	03.22		3100	380	0.25
		Steel castings			
P1.5.C.UT	06.1	Un-alloyed	1400	150	0.25
P2.6.C.UT	06.2	Low-alloy steel (alloying elements ≤ 5%)	1600	200	0.25
P3.0.C.UT	06.3	High-alloy steel (alloying elements > 5%)	1950	200	0.25

The tooth's root width can be calculated by:

$$W_r = M\left(\frac{\pi}{2} - 2\tan\alpha\right) \tag{10.3}$$

where W_r is the root width, mm; M is the module size, mm; and α is the pressure angle, deg.

The flank offset from root of teeth can be calculated by:

$$O = \sqrt{\frac{4M^2}{\cos^2\alpha} - H^2} \tag{10.4}$$

where O is the flank offset from root, mm; and H is the whole depth of tooth space, mm.

Once W_r and O have been computed by using Equations 10.3–10.4, the tooth space area can be calculated by using the following formula:

$$A_{ts} = H(O + W_r) \tag{10.5}$$

where A_{ts} is the tooth space area, mm² (see Example 10.3).

Now, the cutting power in gear milling can be calculated by:

$$P_c = \frac{k_c \cdot A_{ts} \cdot f_r}{10^6} \tag{10.6}$$

where P_c is the cutting power, kW; k_c is the specific cutting force (with m_c factored in); and f_r is the feed rate, mm/s (see Example 10.4).

Once the cutting power has been computed, the cutting torque can be calculated by:

$$T_{torq} = \frac{P_c(9550)}{N} \tag{10.7}$$

where T_{torq} is the torque, N·m; P_c is the cutting power, kW; and N is the cutter's rotational speed, rev/min (see Example 10.5).

10.4 GEAR BROACHING

Broaching machining operation has been introduced to the reader in Chapter 9. *Gear broaching* involves the use of a *broach* that replicates the geometry of the gear tooth to be cut. *Gear broaching* is used to produce internal and external cuts in spur gears as well as in helical gears. Broaching to produce external gear teeth is illustrated in Figure 10.5; here the cutting stroke is in the direction of increasing teeth height of the broach.

In cutting internal gear teeth, an internal (round) broach is generally inserted into a pilot or starting hole which is made on the workpiece before machining (see Figure 10.6). Broaching of internal gears requires proper spacing of teeth which may

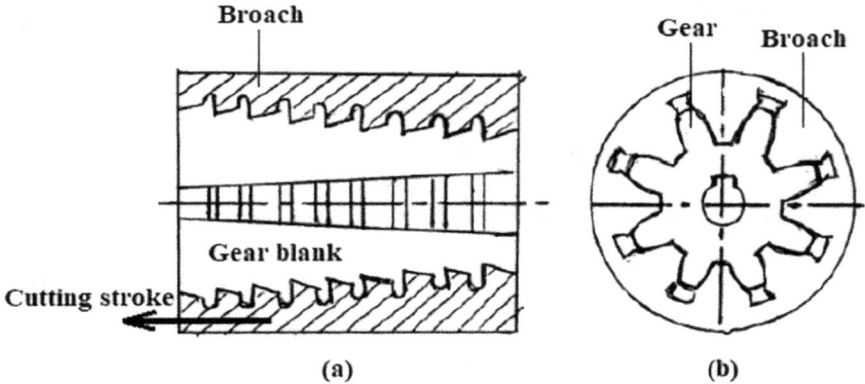

FIGURE 10.5 Broaching external gear teeth (a) and completion of gear broaching operation (b)

FIGURE 10.6 Broaching internal gear teeth

either be equal or unequal; the tooth forms may be symmetrical or asymmetrical. In order to withstand the broaching pressure, the tooth forms must be uniform in the direction of the broach axis, and their surface must be strong enough. Most internal broaching is carried out with pull broaches. A single-pass broaching operation is enough to cut the small internal gears; however, large internal gears are usually broached by using a surface type broach that cuts several teeth with a single pass. A notable example of internal gear is the helical internal gear used in automotive automatic transmission system.

10.5 GEAR SHAPING

10.5.1 Gear Shaping Process

It has been learnt in Section 10.2 that gear shaping is a gear generation process that results in the machined part's geometry dictated by the relative motion of work and cutter. *Gear shaping* is used for cutting external gears, internal gears (or splines), face gears, worm gears, racks, and the like. In gear shaping, the cutter axis and the work (gear blank) axis are parallel to each other. The shaping cutting tool (cutter) is

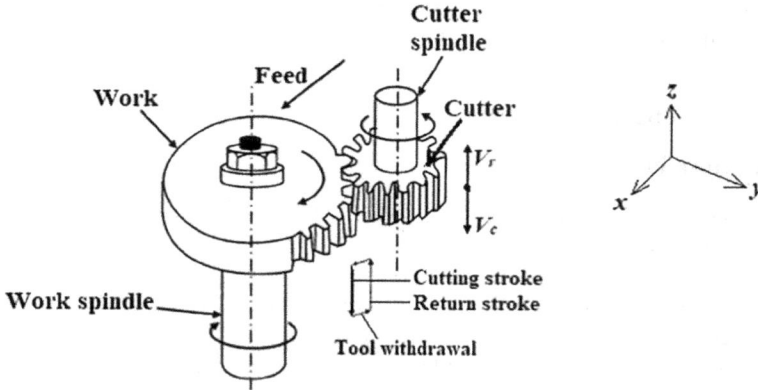

FIGURE 10.7 Gear shaping process. (V_c = cutting stroke velocity, V_r = return stroke velocity)

mounted on a spindle that has three types of motion: (a) axial reciprocating (primary) motion, (b) rotatory generating motion, and (c) feed motion (see Figure 10.7). The work spindle rotates slowly and is synchronized with the cutter spindle. The cutter is fed into the work and moves down for cutting stroke; then it is withdrawn and raised for return stroke. The synchronized rotations of work and tool result in the next tooth cutting, as illustrated in Figure 10.7). In this way, all the teeth on the work (gear blank) are successively cut.

10.5.2 ENGINEERING ANALYSIS OF GEAR SHAPING

In gear shaping, the stroke motion in the z direction (see Figure 10.7) in combination with the rotary motion is performed by the ram of the machine, which contains the so-called lead cartridge. The required stroke length depends on the face width and the starting and overrun path of the shaping-cutter blades. The stroke length for a spur gear blank can be calculated by:

$$l_h = 1.14(b) \tag{10.8}$$

where l_h is the required stroke length, and b is the gear blank face width (see Example 10.6).

The stroke length for a shaper cutter with step cut for a helical gear blank can be computed by (Linke et al., 2016):

$$l_h = \frac{b}{\cos \beta} + (k\, m_n \sin \beta) + \frac{b}{14} \tag{10.9}$$

where β is the helix angle, deg; k is in the range from 0.6 to 1.0, and m_n is the normal module (see Example 10.7).

Engineering analysis of gear shaping has well been studied in depth by researchers; the reader is advised to look into the relevant literature (e.g. Katz, 2017).

10.6 GEAR HOBBING

10.6.1 Gear Hobbing Process

Hobbing is a machining process involves the rotation of both the cutting tool (hob) and the workpiece in a continual timed relationship. Its productivity and versatility render hobbing as one of the most fundamental machining processes for gear manufacturing. Gear hobbing is a generating machining operation *i.e.* the shape of the gear tooth is generated by the combined motions of workpiece (gear blank) and *hob*. The *hob* is a cutting tool that is used to cut the teeth into the workpiece. In order to manufacture a spur gear, the hob is angled equal to the helix angle of the hob. It is important for precise machining that the gear blank and the hob be synchronized in the rotation. While the hob and gear blank are rotating, the hob normally feeds axially across the gear-blank's face at the tooth depth to produce a gear cut (see Figure 10.8). The hob makes successive cuts on the gear blank to generate the gear teeth. For single-start hob, the gear blank will advance one tooth for each revolution of the hob. For a double-start hob, the gear blank is rotated over 2 teeth for each revolution of the hob; for a triple-start hob, the gear blank is rotated over 3 teeth for each revolution of the hob, and so on. Gear hobbing with multi-start hobs results in saving cycle time thereby improving productivity (see Example 10.16).

10.6.2 Hobbing Cutting Tool – *Hob*

A hob is a worm-shaped cutter that has gashes (grooves) cut across it to produce the cutting edges (Figure 10.9a). It is cylindrical in shape with helical cutting teeth. These teeth have gashes that run along the length of the hob and aid in cutting and chip removal. Each cutting tooth is relieved radially to provide chip clearance behind the cutting edge (Figure 10.9b); this "relief" allows the hob face to be sharpened while still maintain the original tooth shape.

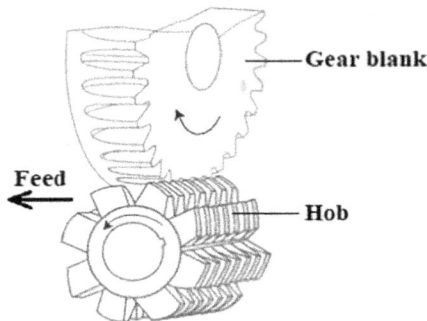

FIGURE 10.8   Gear hobbing operation

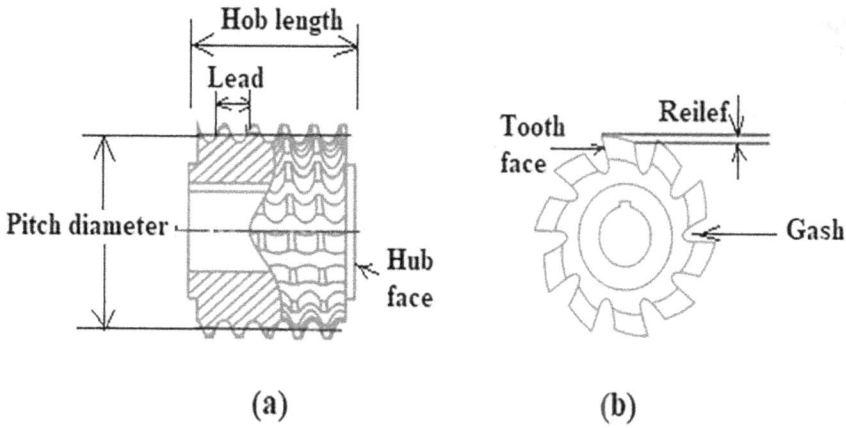

FIGURE 10.9 The hobbing cutting tool: **hob**; (a) front view and (b) side view

10.6.3 HOBBING MACHINE TOOL

A hobbing machine has two non-parallel spindles: one spindle is mounted on a workpiece (e.g. gear blank) and the other spindle is mounted on the hob. Hobbing machines are characterized by the largest "module" or pitch diameter it can generate. There are mainly two types of gear hobbing machines: (a) horizontal hobbing machines and (b) vertical hobbing machines. Horizontal hobbing machines have the gear-blank spindle mounted horizontally; they are usually used for cutting longer workpieces *i.e.* cutting splines at the end of a shaft. In vertical hobbing machine, the gear blank spindle is mounted vertically (see Figure 10.10). Most hobbing machines are vertical hobbers; the vertical hobbing machines can produce all types of pinions and gears/gear wheels.

FIGURE 10.10 A partial view of a gear hobbing machine

10.6.4 Engineering Analysis of Gear Hobbing

A gear hobbing operation may be performed as a single-cut pass or a double-cut pass; the latter involves a cycle consisting of roughing cut followed by a finishing cut. Engineering analyses for both types of passes are presented in the subsequent paragraphs.

10.6.4.1 Engineering Analysis of Single-Cut Pass Gear Hobbing

The cycle time for a single-cut pass gear hobbing process is calculated by (Endoy, 1990):

$$T_c = \frac{Z L_c}{N n_s f} \qquad (10.10)$$

where T_c is the cycle or cutting time, min; Z is the number of gear teeth, L_c is the length of cut, in. or mm; N is the hob's rotational speed, rev/min; n_s is the number of hob starts; and f is the feed, in/rev or mm/rev (see Example 10.8).

The feed, in gear hobbing, strongly depends on the number of hob starts (see sub-section 10.6.1). When working with multi-start hobs involving $Z > 25$, the feed must be reduced to compensate for the increased tooth loading of the hob. Hence, a generalized expression for a hob's feed is given by:

$$f = k f_1 \qquad (10.11)$$

where f is the feed for a multi-start hob, mm/rev; k is the reduction factor; and f_1 is the normal feed for a single-start hob (see Example 10.9). The reduction factor k data for various number of hob starts is given in Table 10.2.

The depth of cut in single-cut pass gear hobbing is given by:

$$d = \frac{OD_{(gear)} - RD_{(gear)}}{2} \qquad (10.12)$$

where d is the depth of cut; $OD_{(gear)}$ is the outside diameter of the gear, and $RD_{(gear)}$ is the root diameter of the gear (see Figure 10.2 and Example 10.10).

10.6.4.2 Engineering Analysis of Double-Cut Pass Gear Hobbing

The engineering analysis of double-cut pass gear hobbing first requires calculation for the hob approach, hob overrun, and the hob travel. The whole tooth depth (H) is given as:

TABLE 10.2
Reduction Factor Data for Various Hob Starts (Endoy, 1990)

Number of hob starts, n_s	1	2	3	4
Reduction factor, k	1	0.67	0.55	0.50

$$H = \frac{OD_g - RD_g}{2} \tag{10.13}$$

where OD_g is the maximum outside diameter of gear, and RD_g is the maximum root diameter of the gear. The depth of cut for roughing (first) cut pass is given by:

$$d_1 = H - a_f \tag{10.14}$$

where d_1 is the depth of cut for roughing (first cut) pass, and a_f is the finishing cut allowance. The depth of cut for finishing cut is $d_2 = a_f$.
   The hob approach for roughing cut for spur gear is given by:

$$A_1 = \sqrt{d_1(D - d_1)} + c \tag{10.15}$$

where A is the hob approach for roughing cut, D is the outside diameter of the hob, and c is the clearance ($c = 0.04$–0.1 in). The hob approach for finishing cut (A_2) for spur gear is given by:

$$A_2 = \sqrt{d_2(D - d_2)} + c \tag{10.16}$$

The significance of Equations 10.13–10.16 is illustrated in Example 10.11.
   The gear addendum is calculated by:

$$AD = \frac{OD_g - PD_g}{2} \tag{10.17}$$

where AD is the gear addendum and PD_g is the pitch diameter of the gear.
   The hob overrun (R) is calculated by:

$$R = \frac{AD \tan \gamma}{\tan \alpha} + c \tag{10.18}$$

where AD is the gear addendum, γ is the hob head swivel angle, α is the pressure angle of the gear, and c is the clearance (see Example 10.12).
   The hob travel in a double-cut pass gear hobbing cycle is computed by:

$$L = A + R + W_f + W_s \tag{10.19}$$

where L is the hob travel, A is the hob approach, R is the hob overrun, W_f is the gear face width, and W_s is the spacer width (see Example 10.13).
   Once the hob travels for the roughing and finishing passes have been computed, the hob's rotational speed and feeds for the two passes are calculated.
   The hob rotational speed for the roughing pass is given by:

$$N_1 = \frac{V_1}{\pi D} \tag{10.20}$$

where N_1 is the rotational speed of the hob in roughing pass, rev/min; V_1 is the peripheral or cutting speed, in/min or mm/min; and D is the hob's outside diameter, in or mm.

Similarly, the hob's rotational speed for the finishing pass (N_2) is given by:

$$N_2 = \frac{V_2}{\pi D} \tag{10.21}$$

The cycle time for a double-cut pass gear hobbing is given by:

$$T_c = \frac{Z\ L_1}{n_s\ N_1\ f_1} + \frac{Z\ L_2}{n_s\ N_2\ f_2} = \frac{Z}{n_s}\left(\frac{L_1}{N_1\ f_1} + \frac{L_2}{N_2 f_2}\right) \tag{10.22}$$

where L_1 is the hob travel in the roughing (1st cut) pass, mm; L_2 is the hob travel in the finishing (2nd cut) pass, mm; N_1 the rotational speed of hob in 1st pass, rev/min; N_2 the rotational speed of hob in 2nd pass, rev/min; f_1 is the feed in 1st pass, mm/rev; and f_2 is the feed in 2nd pass, mm/rev (see Examples 10.14–10.16).

10.7 CALCULATIONS – WORKED EXAMPLES ON GEAR MANUFACTURING

10.7.1 WORKED EXAMPLES ON GEAR MILLING

EXAMPLE 10.1: CALCULATING THE AVERAGE CHIP THICKNESS IN GEAR MILLING

A gear milling cutter has a module of 16 mm with 300 mm outside diameter having 7 effective teeth. The whole depth of tooth space is 34 mm and the pressure angle is 21°. The cutter is used to perform one-pass spur gear milling operation using cutting speed of 160 m/min and feed of 0.5 mm/tooth on a low-alloy steel with BHN 175. Calculate the average chip thickness.

Solution
For spur gear, $a_e = H = 34$ mm, $D_c = 300$ mm, $f_t = 0.5$ mm/tooth, $h_m = ?$
By using Equation 10.1,

$$h_m = \frac{2\,a_e\,f_t}{D_c\cos^{-1}\left(1-\dfrac{2\,a_e}{D_c}\right)} = \frac{2\times34\times0.5}{300\cos^{-1}\left(1-\dfrac{2\times34}{300}\right)}$$

$$= \frac{34}{300\cos^{-1}\left(1-0.226\right)} = \frac{34}{300\times39°}$$

$$h_m = \frac{34}{300\times\left(\dfrac{39.2\,\pi}{180}\right)radian} = \frac{34}{300\times0.685} = 0.165\,\text{mm}$$

The average chip thickness = h_m= 0.165 mm

EXAMPLE 10.2: CALCULATING SPECIFIC CUTTING FORCE WITH m_c FACTORED IN FOR GEAR MILLING

By using the data in Example 10.1, calculate the specific cutting force with m_c factored in.

Solution
Work material: low-alloy steel with BHN 175, $h_m = 0.16$ mm, $k_c = ?$
By reference to Table 10.1, for low-alloy steel (175 BHN), $k_{c1} = 1700$ N/mm^2, $m_c = 0.25$
By using Equation 10.2,

$$k_c = \frac{k_{c1}}{h_m{}^{m_c}} = \frac{1700}{0.16^{0.25}} = \frac{1700}{0.632} = 2689.8 \, \text{N/mm}^2/\text{mm}^{0.25}$$

The specific cutting force with m_c factored in = $k_c = 2689.8$ N/mm^2/mm$^{0.25}$

EXAMPLE 10.3: CALCULATING THE TOOTH SPACE AREA FOR GEAR MILLING

By using the data in Example 10.1, calculate the (a) root width, (b) flank offset from root, and (c) tooth space area for gear milling.

Solution
$M = 16$ mm, $\alpha = 21°$, $H = 34$ mm, $A_{ts} = ?$
a. By using Equation 10.3,

$$W_r = M\left(\frac{\pi}{2} - 2\tan\alpha\right) = 16(1.57 - 2\tan 21) = 16(1.57 - 0.6498) = 12.83 \, \text{mm}$$

b. By using Equation 10.4,

$$O = \sqrt{\frac{4M^2}{\cos^2\alpha} - H^2} = \sqrt{\frac{4\times 16^2}{\cos^2 21} - 34^2} = \sqrt{\frac{1024}{0.8716} - 1156} = \sqrt{18.85} = 4.34 \, \text{mm}$$

c. By using Equation 10.5,

$$A_{ts} = H(O + W_r) = 34(4.34 + 12.83) = 34 \times 17.17 = 583.8 \, \text{mm}^2$$

Root width = 12.83 mm, Flank offset from root = 4.34 mm, Tooth space area = 583.8 mm^2.

EXAMPLE 10.4: CALCULATING THE CUTTING POWER FOR A GEAR MILLING OPERATION

By using the data in Example 10.1, calculate the cutting power for the gear milling operation.

Solution

V = 160,000 mm/min, f_t = 0.5 mm/tooth, D_c = 300 mm, n_t = 7, A_{ts} = 583.8 mm², k_c = 2689.8 ~ 2690 P_c = ?

By using Equation 7.1,

$$N = \frac{v}{\pi D_c} = \frac{160,000}{3.142 \times 300} = 169.74 \, \text{rev / min}$$

By using Equation 7.2,

$$f_r = f_t \, N \, n_t = 0.5 \times 169.74 \times 7 = 594.1 \, \text{mm / min} = \frac{594.1}{60} = 9.90 \, \text{mm/s}$$

By using Equation 10.6,

$$P_c = \frac{k_c \cdot A_{ts} \cdot f_r}{10^6} = \frac{2690 \times 583.8 \times 9.90}{10^6} = 15.55 \, \text{kW}$$

The cutting power in gear milling = P_c = 15.55 kW

EXAMPLE 10.5: CALCULATING THE CUTTING TORQUE IN GEAR MILLING

By using the data in Example 10.4, calculate the cutting torque in the gear milling operation.

Solution

P_c = 15.55 kW, N = 169.74 rev/min, T_{torq} = ?

By using Equation 10.7,

$$T_{torq} = \frac{P_c(9550)}{N} = \frac{15.55 \times 9550}{169.74} = 874.88 \, \text{N–m}$$

The cutting torque in the gear milling operation = 874.88 N–m.

10.7.2 WORKED EXAMPLES ON GEAR SHAPING

EXAMPLE 10.6: CALCULATING THE STROKE LENGTH OF A SHAPER CUTTER FOR CUTTING SPUR GEAR

The face width of a spur gear blank is 20 mm. What should be stroke length of the shaper cutter?

Solution

By using Equation 10.8,

$$l_h = 1.14(b) = 1.14 \times 20 = 22.8 \, \text{mm}$$

The required stroke length of the shaper cutter = 22.8 mm

EXAMPLE 10.7: CALCULATING THE STROKE LENGTH FOR SHAPER CUTTER WITH STEP CUT FOR HELICAL GEAR BLANK

The face width of a helical gear is 15 mm, and its helix angle is 20° R.H. The normal module is 3 mm. Calculate the stroke length for the shaper cutter with step cut for cutting the helical gear.

Solution

$b = 15$ mm, $k = 0.7$, $m_n = 3$, $\beta = 20°$, $l_h = ?$

By using Equation 10.9,

$$l_h = \frac{b}{\cos \beta} + \left(k\, m_n \sin \beta \right) + \frac{b}{14} = \frac{15}{\cos 20} + \left(0.7 \times 3 \times \sin 20 \right) + \frac{15}{14}$$

$$= \frac{15}{0.94} + 0.718 + 1.07 = 17.74 \text{ mm}$$

The stroke length for the shaper cutter with step cut for cutting the helical gear = 17.7 mm

10.7.3 WORKED EXAMPLES ON GEAR HOBBING

EXAMPLE 10.8: COMPUTING THE CYCLE TIME FOR A SINGLE-CUT PASS SINGLE-HOB-START GEAR HOBBING PROCESS

It is required to produce a 20-teeth gear in a single-cut pass with a hob's single-start operation at a cutting speed of 65 m/min. The length of cut is 70 mm, and the feed is 4 mm/rev. The hob's outside diameter is 10 cm. Calculate the cycle time for the hobbing operation.

Solution

$V = 65$ m/min $= 65,000$ mm/min, $Z = 20$ teeth, $L_c = 70$ mm, $f = 4$ mm/rev, $D = 100$ mm, $n_s = 1$

$$\text{Hob's rotational speed} = N = \frac{V}{\pi D} = \frac{65000}{3.142 \times 100} = 207 \text{ rev/min}$$

By using Equation 10.10,

$$T_c = \frac{Z\, L_c}{N\, n_s\, f} = \frac{20 \times 70}{207 \times 1 \times 4} = 1.7 \text{ min}$$

The cycle time for the single-cut pass with a single hob start operation = 1.7 min.

EXAMPLE 10.9: COMPUTING THE CYCLE TIME FOR SINGLE-CUT PASS DOUBLE-HOB START GEAR HOBBING PROCESS

By using the data in Example 10.8, calculate the cycle time for the single-cut pass hobbing operation if the hob used is double start and the number of teeth on the gear is 31.

Solution

$N = 207$ rev/min, $Z = 31$ teeth, $L_c = 70$ mm, $f_1 = 4$ mm/rev, $n_s = 2$, $T_c = ?$

By reference to Table 10.2 for a double-start hob ($n_s = 2$), we get $k = 0.67$

By using Equation 10.11,

$$f = k\,f_1 = 0.67 \times 4 = 2.68\,\text{mm/rev}$$

By using Equation 10.10,

$$T_c = \frac{Z\,L_c}{N\,n_s\,f} = \frac{31 \times 70}{207 \times 2 \times 2.68} = 1.95\,\text{min}$$

EXAMPLE 10.10: CALCULATING THE DEPTH OF CUT FOR A SINGLE-CUT GEAR HOBBING

The outside diameter of a transmission gear is 20 cm, and the root diameter is 19 cm. Calculate the depth of cut for a single-cut hobbing cycle.

Solution

By using Equation 10.12,

$$d = \frac{OD_{(gear)} - RD_{(gear)}}{2} = \frac{20 - 19}{2} = 0.5\,\text{cm} = 5\,\text{mm}$$

The depth of cut for the single-cut hobbing cycle = 5 mm

EXAMPLE 10.11: CALCULATING HOB APPROACHES FOR THE ROUGHING AND FINISHING CUT PASSES

It is required to cut (manufacture) a spur gear with the maximum outside diameter of 9″, and the maximum root diameter of 8.444″. The spur gear has 43 teeth, pressure angle is 20°, face width is 2.5″, and the pitch diameter is 8.811″. The space width is 0.28″. The gear hobbing process design demands single-start double-cut passes cycle by using an HSS hob with an outside diameter of 4.8″, and swivel angle of 3°. The finishing cut allowance is 0.07″. The clearance in the hob approach is 0.045″. The cutting speed in the roughing pass is 240 sfpm and in the finishing pass is 300 sfpm. The feed in the roughing pass is 0.185 in/rev and in the finishing pass is 0.298 in/rev. Calculate the hob approaches for the roughing and finishing cut passes.

Solution

$OD_g = 9.000″$, $RD_g = 8.444″$, $a_f = 0.07″$, $D = 4.8″$, $c = 0.045″$, $A_1 = ?$, $A_2 = ?$

By using Equation 10.13,

$$H = \frac{OD_g - RD_g}{2} = \frac{9.000 - 8.444}{2} = 0.278″$$

By using Equation 10.14,

$$d_1 = H - a_f = 0.278 - 0.07 = 0.208''$$

$$d_2 = a_f = 0.07''$$

By using Equation 10.15,

$$A_1 = \sqrt{d_1(D - d_1)} + c = \sqrt{0.208(4.8 - 0.208)} + 0.045 = \sqrt{0.955} + 0.045 = 1.022''$$

By using Equation 10.16,

$$A_2 = \sqrt{d_2(D - d_2)} + c = \sqrt{0.07(4.8 - 0.07)} + 0.045 = \sqrt{0.331} + 0.045 = 0.62''$$

The hob approach in the roughing cut pass = $A_1 = 1.022'' = 25.96$ mm $\cong 26$ mm
   The hob approach in the finishing cut pass = $A_2 = 0.62'' = 15.75$ mm

EXAMPLE 10.12: CALCULATING HOB OVERRUN FOR A DOUBLE-CUT PASS GEAR HOBBING CYCLE

By using the data in Example 10.11, calculate the hob overrun to perform the gear hobbing operation.

Solution
$OD_g = 9''$, $PD_g = 8.811''$, $\gamma = 3°$, $\alpha = 20°$, $c = 0.045''$, Addendum of gear = AD = ?
   By using Equation 10.17,

$$AD = \frac{OD_g - PD_g}{2} = \frac{9 - 8.811}{2} = 0.094''$$

By using Equation 10.18,

$$R = \frac{AD \tan \gamma}{\tan \alpha} + c = \frac{0.094 \tan 3}{\tan 20} + 0.045'' = 0.01353'' + 0.045'' = 0.0585''$$

The hob overrun for roughing and finishing cuts = $R = 0.0585'' = 1.48$ mm

EXAMPLE 10.13: CALCULATING THE HOB TRAVELS FOR ROUGHING PASS AND FINISHING PASS

By using the data in Example 10.12, calculate the hob travel for the following cutting passes of the gear hobbing cycle: (a) roughing and (b) finishing.

Solution
Gear face width = $W_f = 2.5''$, Spacer width = $W_s = 0.28''$

a. For the roughing cut pass, $A_1 = 1.076''$, $R = 0.0585''$,
By using Equation 10.19 for the roughing cut pass,

$$L_1 = A_1 + R + W_f + W_s = 1.076 + 0.0585 + 2.5 + 0.28 = 3.9145''$$

The hob travel for the roughing cut pass in gear hobbing cycle = $L_1 = 3.9145''$
= 99.43 mm
b. For the finishing cut pass, $A_2 = 0.62''$, $R = 0.0585''$
By using Equation 10.19 for the finishing cut pass,

$$L_2 = A_2 + R + W_f + W_s = 0.62 + 0.0585 + 2.5 + 0.28 = 3.4585''$$

The hob travel for the finishing cut pass in gear hobbing cycle = $L_2 = 3.4585''$
= 87.84 mm

EXAMPLE 10.14: CALCULATING THE CYCLE TIME FOR SINGLE-START DOUBLE-CUT PASSES GEAR-HOBBING PROCESS

By using the data in Examples 10.11–10.13, calculate the cycle time for the double-cut passes gear hobbing process.

Solution
$V_1 = 240$ sfpm = 2880 in/min, $V_2 = 300$ sfpm = 3600 in/min, $f_1 = 0.185$ in/rev, $f_2 = 0.298$ in/rev.
$L_1 = 3.9145''$, $L_2 = 3.4585''$, $Z = 43$, $n_s = 1$, $D = 4.8''$, $Tc = ?$
By using Equation 10.20,

$$N_1 = \frac{V_1}{\pi D} = \frac{2880}{\pi \times 4.8} = 190.96 \, \text{rev/min}$$

By using Equation 10.21,

$$N_2 = \frac{V_2}{\pi D} = \frac{3600}{\pi \times 4.8} = 238.7 \, \text{rev/min}$$

By using Equation 10.22,

$$T_c = \frac{Z}{n_s}\left(\frac{L_1}{N_1 f_1} + \frac{L_2}{N_2 f_2}\right) = \frac{43}{1}\left(\frac{3.9145}{190.96 \times 0.185} + \frac{3.4585}{238.7 \times 0.298}\right)$$
$$= 43(0.1108 + 0.0486) = 6.85 \, \text{min}$$

The cycle time for the double-cut passes gear hobbing process = 6.85 min

EXAMPLE 10.15: CALCULATING THE CYCLE TIME FOR DOUBLE-START DOUBLE-CUT PASSES GEAR HOBBING PROCESS

Repeat Example 10.14 for a double-start gear hobbing process.

Solution

$N_1 = 190.96$ rev/min, $N_2 = 238.7$ rev/min, $L_1 = 3.9145''$, $L_2 = 3.4585''$, $Z = 43$, $n_s = 2$, $D = 4.8''$

By reference to Table 10.2 for a double-start hob ($n_s = 2$), we get $k = 0.67$
For the roughing pass, feed $= k f_1 = 0.67 \times 0.185 = 0.123$ in/rev
For the finishing pass, feed $= k f_2 = 0.67 \times 0.298 = 0.2$ in/rev
By using Equation 10.22,

$$T_c = \frac{Z}{n_s}\left(\frac{L_1}{N_1 f_1} + \frac{L_2}{N_2 f_2}\right) = \frac{43}{2}\left(\frac{3.9145}{190.96\times0.123} + \frac{3.4585}{238.7\times0.2}\right)$$
$$= 21.5(0.166 + 0.072) = 5.12\,\text{min}$$

EXAMPLE 10.16: CALCULATING % SAVING IN CYCLE TIME BY OPTING DOUBLE-START HOBBING

Compare the cycle times of single-start and double-start hobbing as calculated in Examples 10.14–10.15. Compute the percent saving in cycle time by switching from single-start to double-start gear hobbing.

Solution

Cycle time with single-start hobbing = 6.85 min; cycle time with double-start hobbing = 5.12 min,

$$\% \text{ Saving in cycle time} = \frac{6.85 - 5.12}{6.85} \times 100 = 25.25$$

The saving in cycle time = 25.25 %

QUESTIONS AND PROBLEMS

10.1 Draw a sketch of spur gear showing the main nomenclature.
10.2 a. Differentiate between gear forming and gear generation.
      b. Draw the classification chart showing the main gear cutting methods.
10.3 Explain the gear milling process with the aid of a sketch.
10.4 Draw sketches showing the process of broaching external gear teeth.
10.5 Explain gear shaping process with the aid of a sketch.
10.6 Differentiate between the following techniques of gear hobbing:
      a. single-cut pass and double-cut pass,
      b. single-start and multi-start hob,
      c. horizontal hobbing machine and vertical hobbing machine.
P10.7 The face width of a spur gear blank is 22 mm. What should be stroke length of the shaper cutter?
P10.8 The face width of a helical gear is 20 mm, and its helix angle is 17° R.H. The normal module is 2.8 mm. Calculate the stroke length for the shaper cutter with step cut for cutting the helical gear.

P10.9   A gear milling cutter has a module of 13 mm with 280 mm outside diameter
         having 7 effective teeth. The whole depth of tooth space is 30 mm and the
         pressure angle is 20°. The cutter is used to perform one-pass spur gear mill-
         ing operation using cutting speed of 140 m/min and feed of 0.4 mm/tooth on
         a hardened tool steel with BHN 200. Calculate the average chip thickness.

P10.10 By using the data in P10.9, calculate the (a) specific cutting force with m_c
        factored in. (b) root width, (c) flank offset from root, (d) tooth space area, (e)
        cutting power for the gear milling operation, and (f) cutting torque in the
        gear milling operation.

P10.11 The outside diameter of a transmission gear is 22 cm, and the root diameter
        is 21.5 cm. Calculate the depth of cut for a single-cut hobbing cycle.

P10.12 It is required to cut (manufacture) a spur gear with the maximum outside
        diameter of 8.754″, and the maximum root diameter of 8.222″. The spur gear
        has 53 teeth, pressure angle is 18°, face width is 3″, and the pitch diameter is
        8.611″. The space width is 0.24″. The gear hobbing process design demands
        single-start double-cut passes cycle by using an HSS hob with an outside
        diameter of 5″, and swivel angle of 3°. The finishing cut allowance is 0.08″.
        The clearance in the hob approach is 0.042″. The cutting speed in the rough-
        ing pass is 250 sfpm, and in the finishing pass is 310 sfpm. The feed in the
        roughing pass is 0.166 in/rev, and in the finishing pass is 0.234 in/rev.
        Calculate the (a) hob approaches for the roughing and finishing cut passes,
        (b) hob overrun, (c) hob travels for the roughing and finishing passes, and (d)
        cycle time to perform gear hobbing. By using the same data, also calculate
        the cycle time if a triple-start hob were used. Also determine the percent sav-
        ing in cycle time when one switches from single-start to triple-start
        hobbing.

MCQs

10.13   Underline the most appropriate answers for the following statements:
        a. Which gear cutting process ensures improved productivity through
           multi-start option?
           (i) milling, (ii) broaching, (iii) shaping, (iv) hobbing
        b. Which gear cutting process involves tooth-by-tooth rotation (indexing)
           of gear blank?
           (i) milling, (ii) broaching, (iii) shaping, (iv) hobbing
        c. Which gear cutting process has reciprocating speed motion of cutter?
           (i) milling, (ii) broaching, (iii) shaping, (iv) hobbing

REFERENCES

Coromant, S. (2010), *Sandvik Coromant Technical Guide – Turning, Milling, Drilling, Boring,
    Tool Holding*, AB Sandvik Coromant, Sandviken, Sweden.
Endoy, R. (1990), *A Guide to Cycle Time Estimating and Process Planning: Gear Hobbing,
    Planing, and Shaving*, Society of Manufacturing Engineers, Markham, Canada.

Huda, Z. (2020), *Metallurgy for Physicists and Engineers: Fundamentals, Analysis, and Calculations*, CRC Press, Boca Raton, USA.

Katz, A. (2017), *Cutting Mechanics of Gear Shaping Process*, M.A.S Thesis, Department of Mechanical and Mechatronics Engineering, University of Waterloo, Waterloo, Canada.

Lelikov, O.P. (2009) Design of machine elements. In: Grote K.H., Antonsson E. (eds). *Springer Handbook of Mechanical Engineering*, Springer, Berlin, Heidelberg, Germany.

Linke, H., Börner, J., Heß, R. (2016), *Cylindrical Gears: Calculations, Materials, and Manufacturing*, Hanser Publications, Germany.

Marsh, B. (2016), Selecting the Proper Gear Milling Cutter Design for the Machining of High-Quality Parallel Axis, Cylindrical Gears and Splines, *GEARSolutions*, Issue: April 2016.

Part III

Grinding/Abrasive Machining Processes

11 Grinding Operations and Machines

11.1 GRINDING OPERATION AND ITS ADVANTAGES

Grinding is an abrasive machining operation that involves the removal of material by utilizing hard, abrasive particles; which are usually bonded in the form of a grinding wheel. Grinding operation results in improvement of the surface finish of the workpiece by producing small chips. This is why grinding operations are generally performed after work-part geometry has been established by conventional machining. *Surface grinding* is used to produce flat and smooth surfaces by using *surface grinding machines*. Round parts are ground by use of *cylindrical grinding machines* and centerless grinding machines. Form grinders are used to move the work and/or the wheel in various axes to grind surfaces that are precisely contoured. Special grinders used to create and sharpen cutting tools are called tool grinders and cutter grinders, respectively. In surface grinding, the workpiece is held by using a magnetic chuck, and is reciprocated during grinding operation by using a rotating grinding wheel (see Figure 11.1).

There are several advantages of using grinding operation over conventional machining processes; some notable advantages are given as follows.

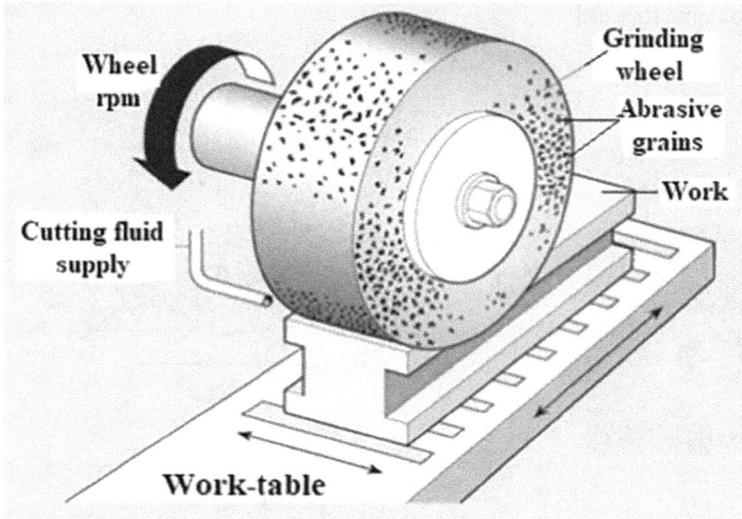

FIGURE 11.1 Surface grinding operation

a. The surface finish obtained by grinding is much superior in quality as compared to the surface finish obtained through a chisel or a file.
b. By using grinding operation, it is conveniently possible to remove metal from a surface that has been hardened – a feature that is missing in conventional machining.
c. As low pressure is required for the grinding operation, it is easier to hold the workpiece by use of simple techniques (*e.g.* a magnetic chuck).
d. The material removal rate (MRR) in grinding, particularly in *creep feed grinding*, is much higher than in conventional machining. In creep feed grinding, the work is fed in a direction opposite to the rotation of grinding wheel rather than in reciprocating motion. This linear non-reciprocating work feed results in larger length of contact between the wheel and the work giving much greater depth of cut and higher MRR (Miao et al., 2020).

11.2 GRINDING WHEEL AND ITS PARAMETERS

11.2.1 GRINDING WHEEL – *GRAINS CUTTING ACTION*

A grinding wheel is a disk comprising abrasive grains and bonded material; the abrasive grains act as cutting teeth during grinding machining operation. Grinding wheels must be precisely balanced for their high rotational speed so as to avoid vibration during the grinding operation (see Figure 11.2a). A grinding wheel is a cutting tool that contains millions of microscopic abrasive grains that are bonded together; here, each abrasive grain acts like a spiky cutting tool. The abrasive grains are held together in the porous structure of the grinding wheel by a bonding polymeric material. When these grains come in contact with the surface to be cut, their sharp edges can cut and remove the material on the surface (see Figure 11.2b). The abrasive grains lose their sharpness after each cutting action; they are regularly removed to allow fresh new grains to come forward.

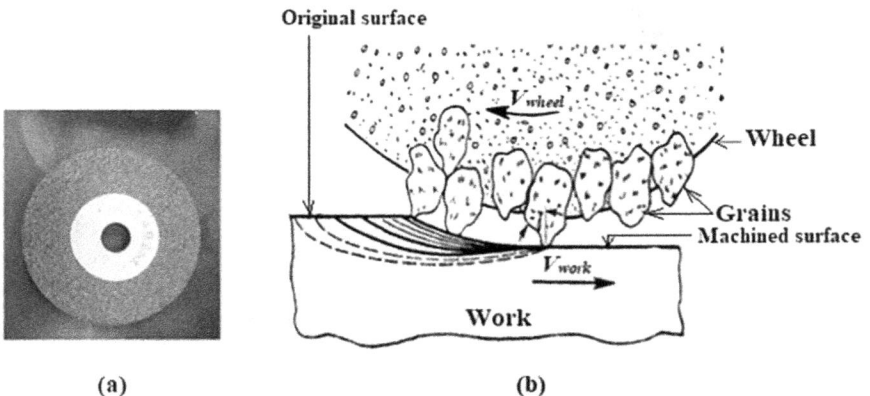

(a)   (b)

FIGURE 11.2   Grinding wheel (a) and cutting action of grains in the wheel (b)

11.2.2 Grinding Wheel Parameters

Grinding wheels are designated/specified by the following *wheel parameters* (Rowe, 2009):

1. Abrasive materials (aluminum oxide, silicon carbide, cubic boron nitride (CBN), diamond),
2. Grit number: 8–600 (8 = very coarse grains; 600 = very fine grains),
3. Wheel grade: A to Z (A = soft; Z = hard),
4. Wheel structure: 1–15 (1 = very dense; 15 = very open), and
5. Bonding material.

The above-mentioned wheel parameters are briefly explained in the following paragraphs.

Abrasive Materials. The grinding-wheel abrasive materials should possess high hardness, wear resistance, toughness, and friability; the latter refers to the capacity to fracture when the cutting edge becomes dull so a new cutting edge is exposed. Aluminum oxide (Al_2O_3) and silicon carbide (SiC) are used as *conventional abrasives*; they are used to grind ferrous and nonferrous alloys with low, medium, and high hardness. *CBN* and diamond are used as *super-abrasives*; they are used to grind very hard materials, such as hardened steels, superalloys, ceramics, cemented carbide, glass, and the like.

Grit Number. Grit number is the inverse of grain size of an abrasive. Coarse grains are specified by grit numbers in the range of 10–60. On the other hand, very fine grains are specified by grit numbers ranging from 320 to 600. The abrasives with high grit numbers (fine grain size) are recommended for excellent surface finish and for cutting hard materials. On the other hand, abrasives with low grit numbers (coarse grain size) are suitable for achieving high MRR and for cutting softer materials.

Table 11.1 presents the grit number ranges for various grain sizes of abrasives.

Wheel Grade. *Wheel grade* determines the strength of the bond that holds the abrasive grains during the grinding operation. It is a function of the amount of bonding material in the grinding wheel. *A-grade* wheels refer to soft wheels that lose grains readily during cutting; they are used for machining hard materials at high cutting speeds with a low feed rate. On the other hand, *Z-grade* wheels are hard wheels

TABLE 11.1
Grit Numbers for Various Grain Sizes of Abrasives

#	Grain Size	Grit number range
1	Very Coarse	8–24
2	Coarse	25–50
3	Medium	60–100
4	Fine	120–220
5	Very fine	230–400
6	Super fine	450–600
7	Superfine for super-abrasives	650–1000

Grains

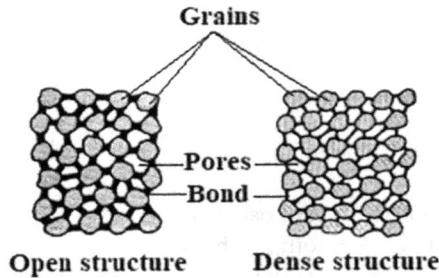

Pores

Bond

Open structure   Dense structure

FIGURE 11.3   The open and dense structures of a grinding wheel

indicating a good grain retain, they are used to machine softer materials at lower cutting speeds with a higher feed rate.

Wheel Structure. The wheel structure determines the spaces between grains in a grinding wheel. A wheel's structure is specified by a number in the range of 1–15. In fact, the volumetric proportion of a wheel is the sum of the volumetric proportions of grains, bond, and voids (pores). A *very open structure* (*e.g.* wheel structure = 1) refers to a large volumetric proportion of pores and small volumetric proportion of grains. A *very dense structure* (*e.g.* wheel structure = 15) indicates large volumetric proportion of grains and small volumetric proportion of pores (see Figure 11.3).

Bonding Material. The grinding wheel bonding material must be able to withstand centrifugal forces and high temperatures resulting from high rotational speeds of the wheel. Another requirement in the bonding material is the resistance to shattering during shock loading of wheel. A bonding material must hold the abrasive grains rigidly in place to effectively accomplish the cutting action; it must also allow the worn grains to be dislodged to expose new sharp grains during grinding operation. Some commonly used bonding materials include vitrified bond (V) (*e.g.* clay), resinoid bond (B) (*e.g.* synthetic resins), silicate bond (S), rubber bond (R), rubber reinforced (RF), and metal bond (M) (for electrically conductive wheels).

Based on the description of grinding wheel parameters, a wheel may be marked by using wheel marking system, as illustrated in Examples 11.1–11.2.

11.3 SURFACE FINISH – *ACHIEVING A GOOD SURFACE FINISH IN GRINDING*

The performance and service life of a component strongly depends on its surface finish/quality. Ideally a surface with a good surface finish should be perfectly flat and smooth; however, this perfectness is not achieved in real-life industrial practice. As a rule of thumb, the better the surface finish/quality of a component, better is its performance and service/fatigue life. Surface finish is the inverse of *surface roughness*; the latter is explained in the following section.

In view of the discussion in the preceding section, the best surface finish can be achieved by controlling the following grinding machining parameters: (a) fine grain size (high grit number) of the abrasive, (b) high rotational speed of the wheel, and (c) dense wheel structure (close to 15). It is interesting to note that the grit number depends on the number of active grains per unit area on the outside periphery of the grinding wheel (see Figure 11.2).

11.4 SURFACE ROUGHNESS/QUALITY

11.4.1 SURFACE QUALITY

It has been emphasized in the preceding section that surface quality plays an important role in the performance of a machined component. There are some important terms in surface geometry/ quality measurement, which include (a) surface texture (surface roughness, waviness, scratches, etc), (b) lay, and (c) flaws. Figure 11.4 illustrates some ways for surface geometry quantification. *Surface texture* refers to all the details that make up a surface. *Lay* is the direction of roughness on a newly machined surface; the roughest profile will be perpendicular to the lay.

It is evident in Figure 11.4 that the surface in Figure 11.4a, being flat and smooth, indicates the best surface quality. On the other hand, the wavy and rough surface (Figure 11.4d) indicates the poorest surface quality.

11.4.2 SURFACE ROUGHNESS – PARAMETERS AND CALCULATION

Surface roughness is the random variation in the (small sized) surface height. A *rough surface* refers to a non-smooth surface indicating a poor surface finish. There are a number of roughness parameters in use, which include arithmetic average (R_a), root mean squared (R_q), maximum valley depth (R_v), and the like (Whitehouse, 2004). The most common parameter of expressing surface roughness is the *arithmetic average (R_a)*, which is the average area per unit length that is off the mean (center line). The arithmetic average (R_a) is also called *center line average (CLA)* with units μm or μin.

In order to calculate the surface roughness using samples at evenly spaced positions, all the small-sized heights at specified positions are measured and plotted to create a surface profile, as shown in Figure 11.5.

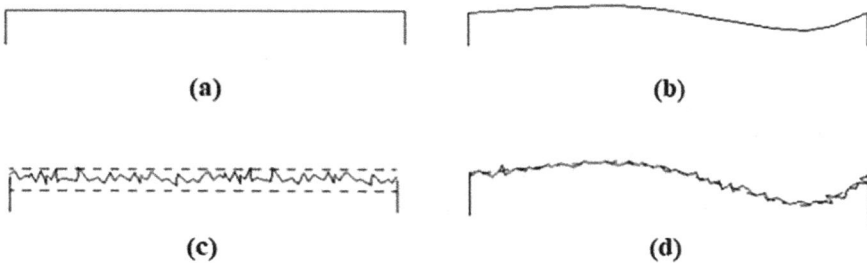

(a)

(b)

(c)

(d)

FIGURE 11.4 Ways of surface geometry quantification: (a) flat and smooth, (b) wavy and smooth, (c) flat and rough, and (d) wavy and rough

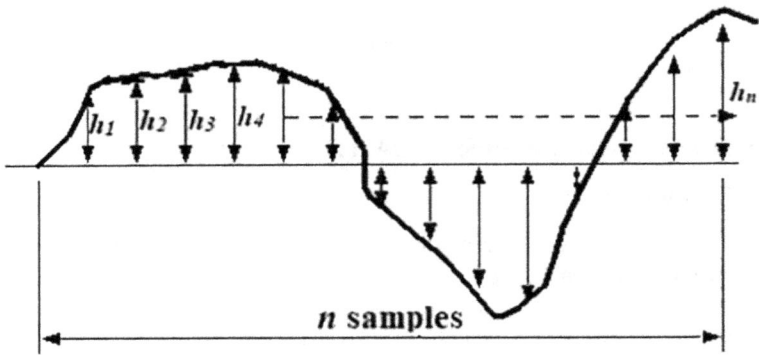

FIGURE 11.5 Surface roughness measurement using heights

By reference to Figure 11.5, the surface roughness R_a can be calculated by dividing the sum of all the small-sized surface heights by the number of samples, as follows:

$$R_a = CLA = \frac{\sum h}{n} = \frac{h_1 + h_2 + h_3 + \ldots + h_n}{n} \tag{11.1}$$

where R_a is the arithmetic average, μm; h_1, h_2, h_3, ...h_n, are the heights at specified points, μm; and n is the number of samples (see Example 11.3).

11.4.3 SURFACE ROUGHNESS MEASUREMENT/TESTING

Surface roughness can be quickly and accurately measured by using a surface roughness tester. A surface roughness tester can directly show the measured mean roughness value (R_a) in micrometers (μm) (see Figure 11.6). Typical roughness testers provide a linear roughness measurement, tracing a mechanical tip along a surface to measure roughness along an arbitrary line. Shorter distances are used to measure finer surfaces, whereas longer lines are traced for rougher surfaces.

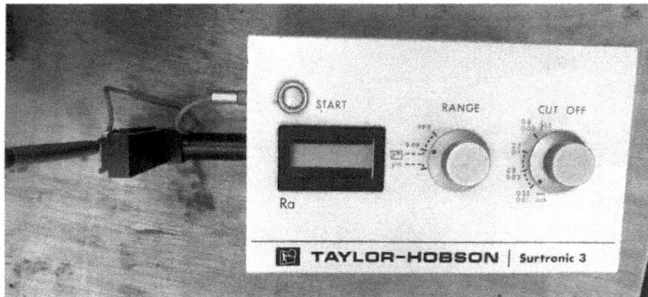

FIGURE 11.6 A surface roughness tester. (*Courtesy*: King Abdulaziz University)

11.5 GRINDING (MACHINING) MACHINES

11.5.1 GRINDING MACHINES AND THEIR TYPES

Grinding machines produce flat, cylindrical, and other surfaces by means of high-speed rotating abrasive wheels (see Section 11.1). A grinding machine consists of a bed with a fixture to guide and hold the workpiece and a power-driven grinding wheel rotating at the required speed. The grinding head can either travel across a fixed workpiece, or the workpiece can be moved while the grinding head stays in a fixed position. In order to avoid overheating, grinding machines incorporate a coolant. There are several types of grinding machines; however, the following three types of machines are generally used: (a) surface grinding machine, (b) cylindrical grinding machine, and (c) centerless grinding machine.

11.5.2 SURFACE GRINDING MACHINES

A surface grinding machine is the most common type of grinding machine found in any machine shop. The surface grinding machine's chuck/worktable has a reciprocating motion once the table has raised the work so that it is slightly deeper within the wheel (see Figure 11.7). As the grinding wheel rotates with its abrasive particles, small amounts of the work's material will be removed each time thereby creating a flat surface. A surface grinding machine is facilitated with a power control unit as well as with hand-wheels to control vertical feed and table traverse feed (see Figure 11.8).

Surface grinding machines are available in four types of structure: (a) horizontal spindle with reciprocating worktable (Figure 11.7), (b) horizontal spindle with rotating worktable, (c) vertical spindle reciprocating worktable, and (d) vertical spindle rotating worktable (Huda, 2017).

11.5.3 CYLINDRICAL GRINDING MACHINES

Cylindrical grinding machines, or cylindrical grinders, are the machine tools that grind the workpiece's external cylindrical surfaces and shoulders. The workpiece in

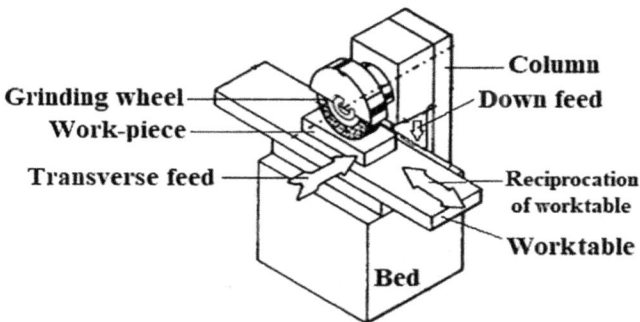

FIGURE 11.7   A schematic of horizontal-spindle surface grinding machine

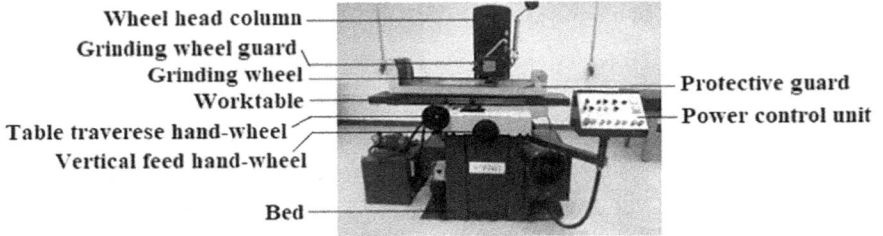

FIGURE 11.8 A surface grinding machine. (*Courtesy*: King Abdulaziz University)

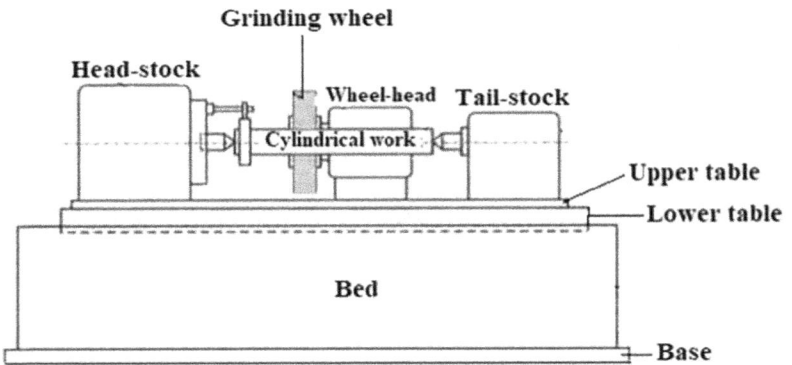

FIGURE 11.9 A schematic illustration of cylindrical grinding machine. (Note: the cylindrical workpiece is held between centers)

cylindrical grinding is held in a chuck, held between centers, or mounted on a face plate in the headstock of the cylindrical grinder (see Figure 11.9). Cylindrical grinding generally involves four actions: (a) the work constantly rotates, (b) the grinding wheel constantly rotates, (c) the grinding wheel is fed toward and away from the work, and (d) either the work or the grinding wheel is traversed with respect to the other. The components that can be ground by cylindrical grinding include spindles, crankshaft bearings, rolls of rolling mills, and the like.

11.5.4 Centerless Grinding Machines

In centerless grinding, the workpiece is not held between centers, but by a rest blade, as shown in Figure 11.10. This technique results in a reduction in work handling time; this is why centerless grinders are high-volume production machines. The cutting speeds in centerless grinding using super-abrasive wheels may be as high as 10,000 m/min (see subsection 11.2.2). Centerless grinding machines are capable of grinding parts with diameters as small as 0.1 mm.

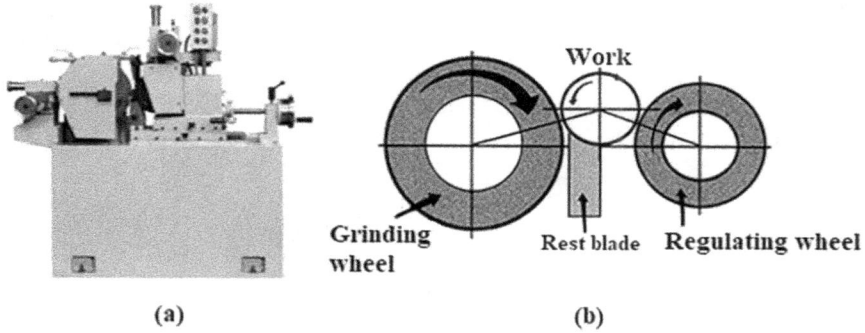

FIGURE 11.10   Centerless grinding machine (a), and centerless grinding operation (b)

11.6 ENGINEERING ANALYSES OF GRINDING OPERATIONS

11.6.1 ENGINEERING ANALYSIS OF SURFACE GRINDING

The variables involved in grinding operation can well be studied by analyzing a surface grinding operation, as shown in Figure 11.11, which illustrates that the grinding wheel removes a layer of metal at a depth of cut d, usually in the range of 0.002–0.05 mm. The work speed (V_w) is generally much smaller than the wheel surface speed (V_{wh}) (see Figure 11.11 and Table 11.2).

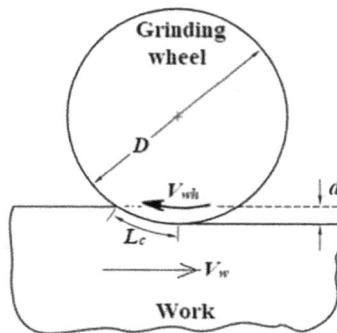

FIGURE 11.11   Machining variables in surface grinding operation

TABLE 11.2
Typical Speed, Feed, and Depth of Cut in Conventional Grinding

Process variable	Wheel surface speed (m/min)	Workspeed (m/min)	Depth of cut (mm)
Value range	1,600–2,800	20–80	0.002–0.05

The selection of cutting speed depends on wheel grade as well as work material (Table 11.2). A high wheel speed may be recommended with super-abrasives wheel, whereas lower cutting speeds are recommended for wheel with conventional abrasives. Once a wheel speed has been selected, the rotational speed of the grinding wheel can be computed by:

$$N = \frac{V_{wh}}{\pi D} \tag{11.2}$$

where V_{wh} is the cutting speed or the surface speed of the wheel in m/min; D is the wheel diameter, m; and N is the rotational speed, rev/min (see Example 11.4).

The MRR is related to the work speed, width, and depth of cut, as follows:

$$MRR = V_w \, w \, d \tag{11.3}$$

where MRR is the material removal rate, mm^3/min; V_w is the workspeed past the wheel, mm/s; w is cross-feed or width of cut, mm; and d is in feed or the depth of cut, mm (see Example 11.5).

The length of contact between the wheel and workpiece, L_c, is given by:

$$L_c = \sqrt{D \, d} \tag{11.4}$$

where L_c is the length of contact or the length of undeformed chip, and d is the depth of cut (see Example 11.6). It may be noted that Equation 11.4 holds good for the condition $V_w \ll V_{wh}$.

The chip formation rate is related to the number of active grains per unit area on the outside periphery of the grinding wheel by:

$$\boldsymbol{n_c} = \boldsymbol{V_{wh}} \, \boldsymbol{w} \, \boldsymbol{n_g} \tag{11.5}$$

where n_c is the number of chips formed per unit time, chips/min; V_{wh} is the surface speed of the wheel, mm/min; w is the width of cut, mm; and n_g is the number of active gains per unit area, grains/mm^2 (see Example 11.7).

The computation of cutting power in grinding requires the use of unit-power (specific energy) data for various materials; the latter is presented in Table 11.3.

Once the MRR has been determined by using Equation 11.3, the cutting power can be computed by reference to Table 11.3 as follows:

$$P_c = P_u (\mathrm{MRR}) \tag{11.6}$$

where P_c is the cutting power, W; P_u is the unit power or specific energy, W-s/mm^3; and MRR is the material removal rate, mm^3/s (see Example 11.8). Once the cutting power is known, the cutting torque ($Torq_c$) can be calculated by using Equation 5.9 (see Example 11.9). By using the so calculated value of $Torq_c$, the cutting force can be computed by:

TABLE 11.3
Unit-Power Data for Surface Grinding (Kalpakjian and Schmid, 2008)

No.	Work Material	Hardness	Unit Power (W-s/mm³)
1	Aluminum	150 BHN	7–27
2	Cast iron	215 BHN	12–60
3	Carbon steel (AISI 1020)	110 BHN	14–68
4	Titanium alloy	300 BHN	16–55
5	Tool steels	67 HR$_C$	18–82

$$F_c = \frac{2\,Torq_c}{D} \qquad (11.7)$$

where F_c is the cutting force, N; $Torq_c$ is the cutting torque, N-m; and D is wheel diameter, m (see Example 11.10).

11.6.2 ENGINEERING ANALYSIS OF CYLINDRICAL GRINDING

For cylindrical grinding, the length of undeformed chip ($L_{c(cyl)}$) is given by:

$$L_{c(cyl)} = \sqrt{\frac{D\,d}{1 + \dfrac{D}{D_{work}}}} \qquad (11.8)$$

where D_{work} is the diameter of workpiece (see Example 11.13).

The traverse feed is the longitudinal feed of the grinding wheel parallel to the axis of the workpiece (see Figure 11.9). The traverse feed is related to the wheel width by (Sharma, 2008):

$$f = n\,w_{wh} \qquad (11.9)$$

where f is the traverse feed, mm/rev; w_{wh} is the width of the wheel, mm; and n is the factor depending on work's diameter (D_w) and rough/finish grinding pass (see Table 11.4 and see Example 11.14).

The time for a cylindrical grinding cut can be calculated by:

TABLE 11.4
The Correction Factor (*n*) Data for Traverse Feed in Cylindrical Grinding

Grinding pass	Rough grinding, $D_w < 20$	Finish grinding, $D_w < 20$	Rough grinding, $D_w \geq 20$	Finish grinding, $D_w \geq 20$
n range	0.3–0.5	0.2–0.4	0.70–0.85	0.2–0.4

$$T_c = \left(\frac{L + 0.5}{f\, N_w} \right) \qquad (11.10)$$

where T_c is the time for a cylindrical grinding cut, min; L is length of workpiece, mm; f is the traverse feed, mm/rev; and N_w is rotational speed of workpiece, rev/min (see Example 11.15).

11.7 CALCULATIONS – *WORKED EXAMPLES ON GRINDING OPERATIONS*

EXAMPLE 11.1: DEFINING A MARKED GRINDING WHEEL BASED ON CONVENTIONAL ABRASIVE

A grinding wheel based on conventional abrasive is marked as follows:

30   C   120   Y   2   RF   XX.

Define each mark of the wheel.

Solution
30 = prefix – manufacturer's symbol for abrasive (optional)
C = Silicon carbide abrasive (A = aluminum oxide, C = silicon carbide)
120 = Grit # 120 = Fine grain size of the abrasive (see Table 11.1)
Y = wheel grade Y = hard wheel (A = soft, Z = hard)
2 = wheel structure 2 = open structure (1 = very open, 15 = very dense)
RF = bonding material is rubber reinforced
XX = manufacturer's trade mark for wheel (optional).

EXAMPLE 11.2: DEFINING A MARKED GRINDING WHEEL BASED ON SUPER-ABRASIVE

A grinding wheel based on super-abrasive is marked as follows:

M   B   550   N   75   V   2.0

Define each mark of the wheel.

Solution
M = Manufacturer's symbol
B = CBN super-abrasive (D = diamond, B = CBN)
550 = grit # 550 = very fine grain size (very coarse = 20, superfine = 1,000)
N = medium hard wheel grade (A = soft, Z = hard)
75 = super-abrasive concentration: 75 = medium-to-high concentration (25 = low, 100 = high)
V = vitrified bond (B = resinoid, M = metal, V = vitrified)
2.0 = 2.0 mm depth of the super-abrasive (1.5 = low depth, 4.5 = high depth) (absence of depth symbol indicates solid diamond or solid CBN).

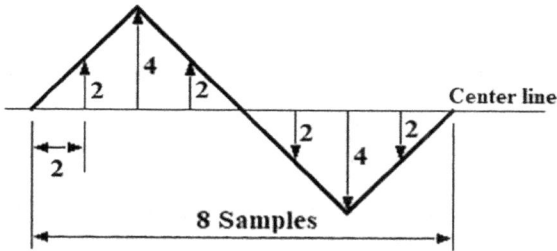

FIGURE E-11.3 Surface profile. (measurements are in μm)

EXAMPLE 11.3: CALCULATING THE SURFACE ROUGHNESS R_a

A surface has a triangular profile as shown in Figure E-11.3. Calculate the surface roughness R_a.

Solution
$n = 8$, $h_1 = 2$, $h_2 = 4$, $h_3 = 2$, $h_4 = 0$, $h_5 = 2$, $h_6 = 4$, $h_7 = 2$, $h_8 = 0$
By using Equation 11.1,

$$R_a = \frac{\Sigma h}{n} = \frac{h_1 + h_2 + h_3 + h_4 + h_5 + h_6 + h_7 + h_8}{8} = \frac{2+4+2+0+2+4+2+0}{8} = 2\,\mu m$$

The surface roughness = arithmetic average = $R_a = 2$ μm

EXAMPLE 11.4: CALCULATING THE WHEEL'S ROTATIONAL SPEED IN SURFACE GRINDING OPERATION

A surface grinding operation is performed at a workspeed of 50 m/min and wheel surface speed of 1,200 m/min. The wheel diameter is 200 mm and the depth of cut is 0.007 mm. The width of cut is 3 mm. Calculate the rotational speed of the wheel.

Solution
$D = 200$ mm $= 0.2$ m, $V_{wh} = 1{,}200$ m/min, $N = ?$
By using Equation 11.2,

$$\text{Grinding wheel's rotational speed} = N = \frac{V_{wh}}{\pi D} = \frac{1{,}200}{\pi \times 0.2} = 1{,}909.6\,\text{rev}/\text{min}$$

EXAMPLE 11.5: CALCULATING THE MRR IN A SURFACE GRINDING OPERATION

By using the data in Example 11.4, calculate the MRR in the operation.

Solution
$V_w = 50$ m/min $= 50{,}000$ mm/min, Width of cut $= w = 3$ mm, $d = 0.007$ mm, $MRR = ?$

By using Equation 11.3,

$$MRR = V_w\,w\,d = 50,000 \times 3 \times 7 \times 10^{-3} = 1,050\,\text{mm}^3/\text{min}$$

The MRR = 1,050 mm³/min = 1.05 cm³/min.

EXAMPLE 11.6: CALCULATING THE LENGTH OF UNDEFORMED CHIP IN SURFACE GRINDING

By using the data in Example 11.4, calculate the undeformed chip length in the operation.

Solution
$D = 200$ mm, $d = 0.007$ mm, $L_c = ?$
By using Equation 11.4,

$$L_c = \sqrt{D\,d} = \sqrt{200 \times 0.007} = \sqrt{1.4} = 1.18\,\text{mm}$$

The undeformed chip length in the grinding operation = 1.18 mm

EXAMPLE 11.7: CALCULATING THE NUMBER OF CHIPS FORMED/MINUTE IN SURFACE GRINDING

A grinding wheel's surface contains 0.8 active grains/mm². The grinding wheel is used to perform surface grinding operation at a wheel speed of 1,500 m/min. The depth of cut is 0.01 mm and the width of cut is 4.5 mm. Calculate the number of chips formed per minute.

Solution
$V_{wh} = 1,500$ m/min $= 15 \times 10^5$ mm/min, $w = 4.5$ mm, $d = 0.01$ mm, $n_g = 0.8$ active grains/mm²
By using Equation 11.5,

Number of chips formed $= n_c = V_{wh}\,w\,n_g = 15 \times 10^5 \times 4.5 \times 0.8 = 54 \times 10^5$ chips/min.

EXAMPLE 11.8: CALCULATING THE CUTTING POWER IN SURFACE GRINDING OF TITANIUM ALLOY

A surface grinding operation is performed to machine a titanium alloy (BHN: 300) workpiece at a workspeed of 50 m/min. The depth of cut is 0.007 mm, and the width of cut is 3 mm. The 200-mm-diameter wheel rotates at 3,000 rev/min. Calculate the cutting power in the operation.

Solution
$V_w = 50,000$ mm/min, $d = 0.007$ mm, $w = 3$ mm, $P_u = 30$ W-s/mm³ (see Table 11.3), $P_c = ?$

By using Equation 11.3,

$$MRR = V_w\,w\,d = 50{,}000 \times 3 \times 7 \times 10^{-3} = 1{,}050\,\text{mm}^3/\text{min} = 17.5\,\text{mm}^3/\text{s}$$

By using Equation 11.6,

$$P_c = P_u(\text{MRR}) = 30\frac{W \cdot s}{\text{mm}^3} \times 17.5\frac{\text{mm}^3}{s} = 525\,\text{W}$$

The cutting power in the grinding operation $= P_c = 525$ W

EXAMPLE 11.9: CALCULATING THE CUTTING TORQUE IN GRINDING OPERATION

By using the data in Example 11.8, calculate the cutting torque in the grinding operation.

Solution
$P_c = 525$ W, N = 3,000 rev/min, $Torq_c = $?

$$N_{(rad/s)} = 2\pi N = 2 \times 3.142 \times 3{,}000 = 18{,}852\,\text{rad/min} = 314.2\,\text{rad/s}$$

By using Equation 5.9.

$$Torq_c = \frac{P_c}{N_{(rad/s)}} = \frac{525}{314.2} = 1.67\,\text{N} - \text{m}$$

The cutting torque $= Torq_c = 1.67$ N-m.

EXAMPLE 11.10: CALCULATING THE CUTTING FORCE IN GRINDING OPERATION

By using the data in Equation 11.9, calculate the cutting force in the grinding operation.
$Torq_c = 1.67$ N-m, $D = 200$ mm $= 0.2$ m, $F_c = $?
By using Equation 11.7,

$$F_c = \frac{2\,Torq_c}{D} = \frac{2 \times 1.67}{0.2} = 16.7\,\text{N}$$

The cutting force $= 16.7$ N

EXAMPLE 11.11: CALCULATING THE CUTTING POWER, CUTTING TORQUE, AND CUTTING FORCE IN SURFACE GRINDING OF AN ALUMINUM ALLOY

A surface grinding operation is performed to machine an aluminum alloy (BHN: 150) at a workspeed of 40 m/min. The depth of cut is 0.02 mm, and the width of cut is 2.8 mm. The 200-mm-diameter wheel rotates at 2,800 rev/min. Calculate the:

(a) cutting power, (b) cutting torque, and (c) cutting force in the grinding operation.

Solution
$V_w = 40,000$ mm/min, $d = 0.02$ mm, $w = 2.8$ mm, $P_u = 18$ W-s/mm³ (see Table 11.3), $P_c = ?$
    a. By using Equation 11.3,

$$MRR = V_w\, w\, d = 40,000 \times 2.8 \times 0.02 = 2,240\, mm^3/min = 37.33\, mm^3/s$$

By using Equation 11.6,

$$\text{The cutting power} = P_c = P_u(MRR) = 18\frac{W \cdot s}{mm^3} \times 37.33\frac{mm^3}{s} = 672\, W$$

    b. $P_c = 672$ W, $N = 2,800$ rev/min, $Torq_c = ?$

$$N_{(rad/s)} = 2\pi N = 2 \times 3.142 \times 2,800 = 17,595\, rad/min = 293.25\, rad/s$$

By using Equation 5.9.

$$\text{The cutting torque} = Torq_c = \frac{P_c}{N_{(rad/s)}} = \frac{672}{293.25} = 2.29\, N-m$$

    c. $Torq_c = 2.29$ N-m, $D = 200$ mm $= 0.2$ m, $F_c = ?$
    By using Equation 11.7,

$$\text{The cutting force} = F_c = \frac{2\, Torq_c}{D} = \frac{2 \times 2.29}{0.2} = 22.9 \cong 23\, N$$

EXAMPLE 11.12: CALCULATING THE CUTTING SPEED AND LENGTH OF CONTACT BY USING A SKETCH

By using the machining data in Figure E-11.12, calculate (a) the wheel's surface speed and (b) the length of undeformed chip in the surface grinding operation.

Solution
$D = 200$ mm, $d = 0.005$ mm, $N = 1,800$ rev/min, $V_{wh} = ?$ $L_c = ?$
    a. By using the modified form of Equation 11.2,

$$V_{wh} = \pi D N = \pi \times 200 \times 1,800 = 1131,120\, mm/min = 1131.12\, m/min$$

The wheel's surface speed $= 1131.12$ m/min
    b. By using Equation 11.4,

$$\text{The length of undeformed chip} = L_c = \sqrt{Dd} = \sqrt{200 \times 0.005} = 1\, mm$$

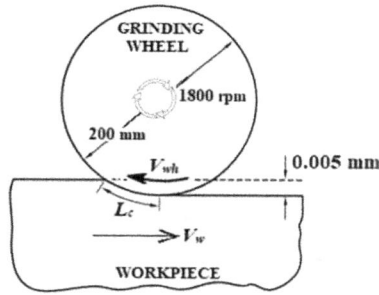

FIGURE E-11.12 The sketch of surface grinding operation showing machining data

The reader is advised to look at the fascinating cover design image of this book.

EXAMPLE 11.13: CALCULATING THE LENGTH OF UNDEFORMED CHIP IN CYLINDRICAL GRINDING

A cylindrical grinding is performed by using a wheel diameter of 240-mm-diameter grinding wheel. The diameter of workpiece is 150 mm, and the depth of cut is 0.03 mm. Calculate the length of undeformed chip for the grinding operation.

Solution
$D = 240$ mm, $D_{work} = 150$ mm, $d = 0.03$ mm, $L_{c(cyl)} = ?$
   By using Equation 11.8,

$$L_{c(cyl)} = \sqrt{\frac{D\,d}{1 + \dfrac{D}{D_{work}}}} = \sqrt{\frac{240 \times 0.03}{1 + \dfrac{240}{150}}} = \sqrt{\frac{7.2}{2.6}} = 1.66\,\text{mm}$$

   The length of undeformed chip for the cylindrical grinding operation = 1.66 mm

EXAMPLE 11.14: CALCULATING THE TRAVERSE FEED IN ROUGH AND FINE CYLINDRICAL GRINDING

A 3-mm-wide grinding wheel is used to perform cylindrical grinding operation on a 300-mm long 25-mm-diameter workpiece rotating at 200 rev/min. Calculate the traverse feed in (a) rough grinding and (b) fine grinding.

Solution
$w_{wh} = 3$ mm, $D_w > 20$ mm, n (rough) = 0.8, n (finish) = 0.3 (see Table 11.4)
   a. By using Equation 11.9,

   Traverse feed in rough grinding $= f = n\,w_{wh} = 0.8 \times 3 = 2.4\,\text{mm/rev.}$

   b. Traverse feed in finish grinding $= f = n\,w_{wh} = 0.3 \times 3 = 0.9$ mm/rev.

EXAMPLE 11.15: CALCULATING THE TIME IN ROUGH AND FINE CYLINDRICAL GRINDING CUTS

By using the data in Example 11.14, calculate the time for a cylindrical grinding cut in (a) rough grinding and (b) finish grinding.

Solution
N_w = 200 rev/min, L = 300 mm, f (rough) = 2.4 mm/rev, f (finish) = 0.9 mm/rev
   a. By using Equation 11.10,

$$\text{The time in rough grinding cut} = T_c = \left(\frac{L+0.5}{f\,N_w}\right) = \left(\frac{300+0.5}{2.4\times 200}\right) = 0.626\,\text{min}$$

$$\text{b. The time in finish grinding cut} = T_c = \left(\frac{L+0.5}{f\,N_w}\right) = \left(\frac{300+0.5}{0.9\times 200}\right) = 1.67\,\text{min}$$

QUESTIONS AND PROBLEMS

11.1   Highlight the advantages of grinding machining, including those of *creep feed grinding*.
11.2   Diagrammatically illustrate the cutting action of grains in surface grinding operation.
11.3   a. List grinding wheel parameters.
       b. Explain wheel structure with the aid of sketches.
11.4   a. Explain the meaning and technological importance of *surface finish*.
       b. How can a good surface finish be achieved by surface grinding?
11.5   a. Define the terms: *surface texture*, *surface roughness*, and *lay*.
       b. Diagrammatically illustrate the ways of surface geometry quantification.
       c. How is surface roughness R_a measured by using a surface roughness tester?
11.6   Explain the working of a surface grinding machine with the aid of a sketch.
11.7   Differentiate between (a) surface grinding, (b) cylindrical grinding, and (c) centerless grinding.
P11.8 A grinding wheel based on conventional abrasive is marked as follows:

30   A   500   B   8   V   XX.

Define each mark of the wheel.

P11.9 Calculate the surface roughness R_a for the surface profile shown in Figure P11.9.
P11.10 A surface grinding operation is performed at a workspeed of 45 m/min and wheel surface speed of 1,300 m/min. The wheel diameter is 180 mm and the depth of cut is 0.013 mm. The width of cut is 2.5 mm. Calculate the (a) rotational speed of the wheel, (b) MRR, and (c) length of undeformed chip in the grinding operation.

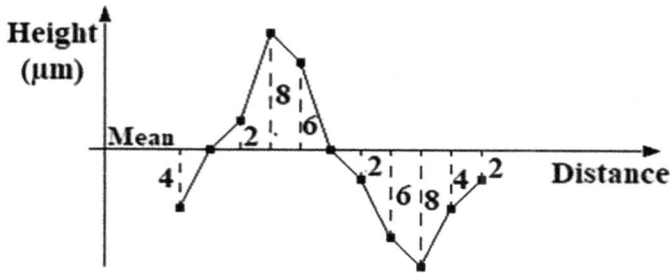

FIGURE P11.9   The profile of a surface

P11.11   A cylindrical grinding is operation performed by using a wheel diameter of 220-mm-diameter grinding wheel. The diameter of workpiece is 180 mm, and the depth of cut is 0.02 mm. Calculate the length of undeformed chip for the grinding operation.

P11.12   A grinding wheel based on super-abrasive is marked as follows:

$$M \quad D \quad 80 \quad B \quad 100 \quad M \quad 2.0$$

Define each mark of the wheel.

P11.13   A grinding wheel's surface contains 1.2 active grains/mm^2. The grinding wheel is used to perform surface grinding operation at a wheel speed of 1,300 m/min. The depth of cut is 0.03 mm and the width of cut is 3.8 mm. Calculate the number of chips formed/min.

P11.14   A surface grinding operation is performed to machine a carbon steel workpiece (BHN: 110) at a workspeed of 60 m/min. The depth of cut is 0.02 mm, and the width of cut is 4 mm. The 240-mm-diameter wheel rotates at 3,200 rev/min. Calculate the (a) cutting power, (b) cutting torque, and (c) cutting force in the grinding operation.

P11.15   (MCQs). Underline the most appropriate answers:

   a. In which type of grinding machine is the workpiece held between centers?
      (i) Surface grinding, (ii) cylindrical grinding, (iii) centerless grinding, (iv) none
   b. In which type of grinding can a workpiece with very small diameter be machined?
      (i) Surface grinding, (ii) cylindrical grinding, (iii) centerless grinding, (iv) none
   c. Which type of grinding machine is most commonly used in machine shops?
      (i) Surface grinding, (ii) cylindrical grinding, (iii) centerless grinding, (iv) all equally
   d. Which grinding machining parameters/variables result in the best surface finish?

         (i) high wheel rotational speed and coarse abrasive grain size, (ii) low rotational speed and fine grain size, (iii) high rotational speed and fine grain size, (iv) low rotational speed and coarse grain size.

P11.16   A 2.5-mm-wide grinding wheel is used to perform cylindrical grinding operation on a 220-mm long 15-mm-diameter workpiece rotating at 170 rev/min. Calculate the (a) traverse feeds in rough grinding and in fine grinding, (b) times in rough grinding cut, and in finish grinding cut.

REFERENCES

Huda, Z. (2017), *Materials Processing for Engineering Manufacture*, Trans Tech Publications, Pfaffikon, Switzerland.

Kalpakjian, S. and Schmid, S.R. (2008), *Manufacturing Processes for Engineering Materials* (5th Edition), Prentice Hall – Pearson Education South Asia Pvt Ltd, Singapore.

Miao, Q., Ding, W., Kuang, W., Yang, C. (2020), Comparison on grindability and surface integrity in creep feed grinding of GH4169, K403, DZ408, and DD6 nickel-base superalloys, *Journal of Manufacturing Processes*, 49, 175–186.

Rowe, W.B. (2009), *Principles of Modern Grinding Technology*, Elsevier Science Publications, Waltham, MA, USA.

Sharma, P.C. (2008), *Manufacturing Technology-II*, S. Chand and Company Pvt. Ltd, New Delhi.

Whitehouse, D. (2004), *Surface and Their Measurement*, Butterworth-Heinemann Elsevier Ltd, Oxford, UK.

12 Abrasive Finishing Machining Operations

12.1 ABRASIVE FINISHING MACHINING AND THEIR APPLICATIONS

Abrasive finishing machining processes involve finishing grinding operations to achieve superior surface finish. Grinding machining operations can only produce surface finishes with surface roughness R_a as low as 0.2 μm. However, in many industrial applications, much superior surface finishes with R_a in the range of 0.013–0.2 μm are desired; this objective is achieved by abrasive finishing processes (see Table 12.1). Many automotive and other machine elements can be finished by abrasive finishing processes to obtain attractive surface finish; these components include gun barrels, cylinder bores of internal combustion engines, cylinder liners, crankshafts, hydraulic cylinders, bearing surfaces, gages, optical lenses, and the like.

There are several types of abrasive finishing machining operations; these include (a) honing, (b) lapping, (c) superfinishing, (d) polishing, and the like. Each of these finishing operations is explained in the subsequent sections.

12.2 HONING AND IT'S ENGINEERING ANALYSIS

12.2.1 HONING OPERATION AND ITS APPLICATIONS

Honing is an abrasive finishing operation that produces functional surface finish by use of a honing tool. The honing tool consists of a set of bonded abrasive sticks that are equally spaced around the tool's periphery. The grit numbers of abrasive sticks in a honing tool lie in the range of 300–600. In honing, the honing tool is moved through the cylinder bore, while the abrasive sticks are highly pressed outwards (usually exercised by small springs) against the bore surface (see Figure 12.1). The honing tool is given a complex rotational and reciprocating axial motions, which combine to produce a crosshatched pattern with topographic form of the surface roughness (see Figure 12.2). This crosshatched pattern may either be a smooth surface or a rough topography with defined oil pockets to retain the lubrication. A cutting fluid must be used in honing for cooling, lubrication, and the removal of chip.

Honing operation can produce good surface finishes with R_a as low as 0.1 μm. Honed surfaces are always in good tolerances. Typical components that are subjected to the honing operation include cylinder bores of internal combustion engines, crankshafts, gun barrel, and the like.

TABLE 12.1
The Surface Roughness Obtained by Some Abrasive Finish Machining Operations

No.	Abrasive finishing operation	Usual part shape	Surface roughness (μm)
1.	Honing	Round hole	0.1–0.7
2.	Lapping	Flat, near flat	0.025–0.3
3.	Superfinishing	Flat, cylindrical	0.01–0.025
4.	Polishing	Misc. shape	0.025–0.4

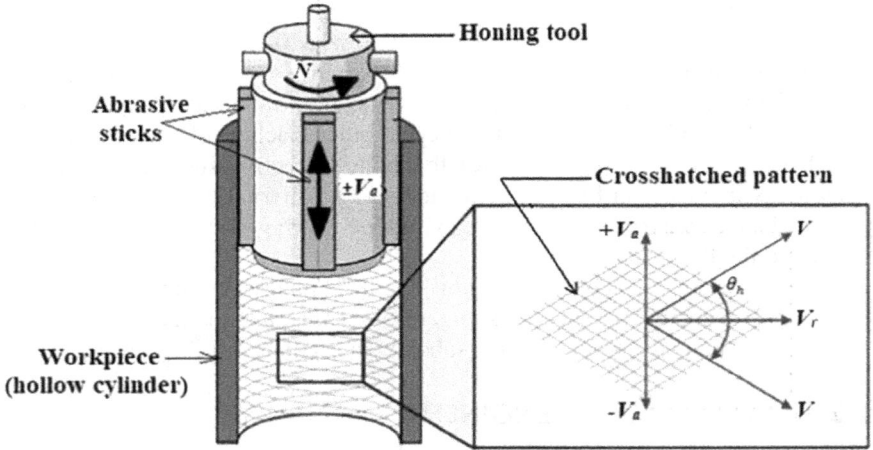

FIGURE 12.1 Honing operation. (V_a is the reciprocating speed, V_r is the peripheral speed, V is the resultant cutting speed, and N is the tool's rotational speed)

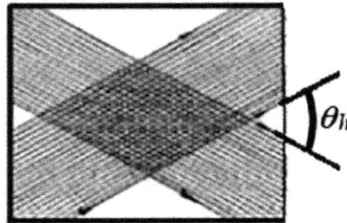

FIGURE 12.2 The crosshead pattern in a honed surface. (θ_h = crosshatch angle)

12.2.2 MATHEMATICAL MODELING OF HONING

The reciprocating speed of the honing tool can be determined by the formula (Zhang, et al., 2012):

$$V_a = 2LS \tag{12.1}$$

where V_a is the reciprocating or stroke speed, m/min; L is the stroke length, m; and S is the number of strokes per minute, strokes/min (see Example 12.1).

The surface speed of the rotating honing tool can be calculated by:

$$V_r = \pi D N \tag{12.2}$$

where V_r is the surface speed of the tool, m/min; D is the honing tool's diameter, m; and N is the rotational speed of the tool, rev/min (see Example 12.2).

The resultant of the reciprocating speed and the surface speed of the tool can be computed by:

$$V = \sqrt{V_a^2 + V_r^2} \tag{12.3}$$

where V is the resultant of the reciprocating speed (V_a) and the surface speed (V_r) of the tool, m/min (see Example 12.3). In order to create the desired crosshead pattern with the right honing angle, there must be a balance between the speeds V_a and V_r (see Figures 12.1–12.2).

The crosshatch angle can be calculated by (Edberg and Landqvist, 2015):

$$\theta_h = 2\tan^{-1}\left(\frac{V_a}{V_r}\right) \tag{12.4}$$

where θ_h is the crosshatch angle, deg.; V_a is the reciprocating or stroke speed, m/min; and V_r is the surface, tangential, or peripheral speed of the honing tool, m/min (see Example 12.4).

12.3 LAPPING – *OPERATION, ADVANTAGES, APPLICATIONS, AND ANALYSIS*

12.3.1 LAPPING OPERATION

Lapping is an abrasive finishing operation that involves the use of fluid suspension of very fine abrasive particles between workpiece and the lapping tool – *lap*. The *lap* may be made of a soft material (*e.g.* copper, lead, wood, etc); it is the inverse of the desired shape of the workpiece. The abrasive particles may be either alumina (Al_2O_3) or silicon carbide (SiC) (grit no. range: 300–600). The fluid suspension is prepared as a *lapping slurry* – a mixture of oil-based fluid and abrasive grains; the slurry has a chalky-paste appearance. In lapping, the *lap* is pressed against the work and is either rotated or moved back and forth over the surface (see Figure 12.3). The lapping pressure is from 0.01 to 0.7 MPa. The abrasive particles, during lapping, roll and move

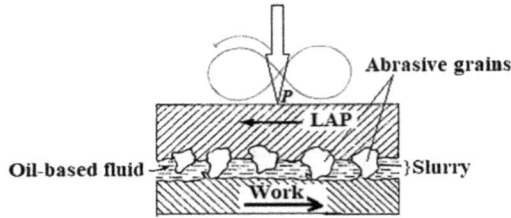

FIGURE 12.3   Lapping operation. (p = lapping force)

freely in all directions, thereby making the surface smooth by uniformly removing a small amount of material (lapping machining allowance = 0.005–0.01 mm).

12.3.2 ADVANTAGES AND APPLICATIONS OF LAPPING

There are several distinct advantages of lapping. No or very little heat is generated in lapping. Lapped surfaces are flat and smooth with the surface roughness R_a in the range of 0.025–0.3 μm. Lapped surfaces have reduced peaks and valleys resulting in the maximum bearing area between mating surfaces, which ensures tight seating of seals thereby eliminating the need of gaskets in automotive and other engineering components. A lapped component has improved fatigue life and wear resistance owing to a good surface finish. The dimensional accuracy of a lapped component is excellent. Notable applications of lapped components include optical lenses, ball bearings, roller bearings, jigs' bushes, surface plates, injectors of diesel engines, and the like.

12.3.3 MATHEMATICAL MODELING OF LAPPING OPERATION

The surface finish of a lapped surface strongly depends on the abrasive concentration used in the slurry. The abrasive concentration C in the slurry can be calculated by:

$$C = \frac{m_a}{m_a + m_l} \tag{12.5}$$

where m_a is the mass of abrasive, g; and m_l is the mass of carrier fluid, g (see Example 12.5).

The material removal rate (MRR) in lapping depends on the height of the material removed (h) and the machining time (see Figure 12.4).

The material removal rate in lapping is generally calculated by (Zhu, 2011; Preston, 1927):

$$\text{MRR}_{lapping} = K\,v\,p \tag{12.6}$$

where $\text{MRR}_{lapping}$ is the material removal rate in lapping, $(\mu m)^3/s$; K is the Preston's coefficient, $(\mu m)^3/\text{N-m}$; v is the lapping speed, m/s; and p is the lapping force, N (see Example 12.7).

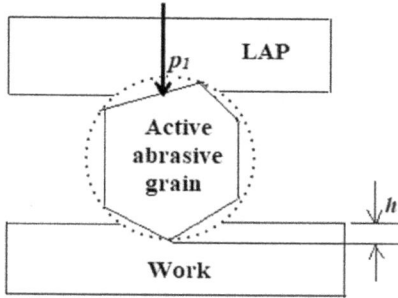

FIGURE 12.4   The cutting action of an active abrasive grain in lapping. (h = the height of the material removed, p_1 = the partial pressure on the grain)

Recently, a new method of computing MRR in lapping has been reported, as follows (Deaconwscu and Deaconwscu, 2020):

$$\text{MRR}_{lapping} = \frac{h}{T_m} \qquad (12.7)$$

where $\text{MRR}_{lapping}$ is the material removal rate in lapping, μm/min; h is the height of the material removed, μm; and T_m is the machining (lapping) time, min (see Example 12.8).

12.4 SUPERFINISHING

Superfinishing is an abrasive finishing operation that involves simultaneous reciprocating motion of a single-bonded abrasive stone against a rotary motion of workpiece (Figure 12.5). An advanced cutting fluid is used in superfinishing that builds up a thin lubricant film between the tool and work surface. Superfinishing produces excellent mirror-like surface finishes with extremely low R_a in the range of 0.01–0.025 μm. Some superfinishing process parameters are presented in Table 12.2.

The machining time in superfinishing operation can be calculated by (El-Hofy, 2018):

$$T_{sf} = \frac{L\,a_m}{f\,N\,d} \qquad (12.8)$$

where T_{sf} is the superfinishing time, min; a_m is the machining allowance, mm; L is the length of the work surface, mm; d is the depth of cut, mm; f is the longitudinal feed, mm/rev; and N is the work rotational speed, rev/min (see Example 12.10).

12.5 POLISHING

Polishing is an abrasive finishing operation aimed at enhancing the appearance of an item, prevent contamination of instruments, remove corrosion effects, create a reflective surface, or prevent rusting in pipes. In metallographic practice, a polishing

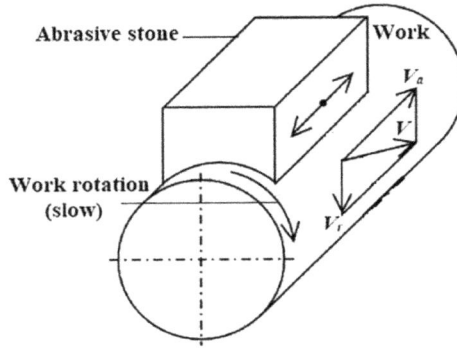

FIGURE 12.5 Superfinishing operation. (V_a = reciprocating velocity of the stone, V_r = peripheral velocity of the work, V = resultant velocity) (see Example 12.8)

TABLE 12.2
Some Superfinishing Process Parameters

Process Parameter	Frequency of Oscillation	Stroke Length	Reciprocating Speed of the Stone	Peripheral Speed of the Work
Value range	200–900 strokes/min	2–4 mm	6–9 m/min	9–100 m/min

operation involves the removal of the remaining scratches on the ground specimen so as to produce a smooth mirror-like surface for examination of a metal's microstructure under a microscope (Huda, 2020). Polishing is accomplished by employing a polishing machine with rotating polishing disk that are covered with soft cloth impregnated with abrasive particles (*e.g.* high alumina suspensions, diamond paste, etc.) and an oily or water lubricant (see Figure 12.6a).

In metallographic polishing using diamond paste as the polishing medium, first a coarser 6-μm-diameter particles polish is applied, and then a finer 1-μm particles polish is used. After each polishing step, the metallographic specimen should be thoroughly washed with warm soapy water, and finally an alcohol rinse should be used followed by drying using a hot air drier. The metallographic polishing produces a smooth mirror-like shining surface (Figure 12.6b).

12.6 CALCULATIONS – *WORKED EXAMPLES ON ABRASIVE FINISHING OPERATIONS*

EXAMPLE 12.1: CALCULATING THE RECIPROCATING SPEED OF THE HONING TOOL

The stroke length of a honing tool, during a honing operation, is 500 mm. Calculate the reciprocating speed of the honing tool if the tool reciprocates 15 times during the operation.

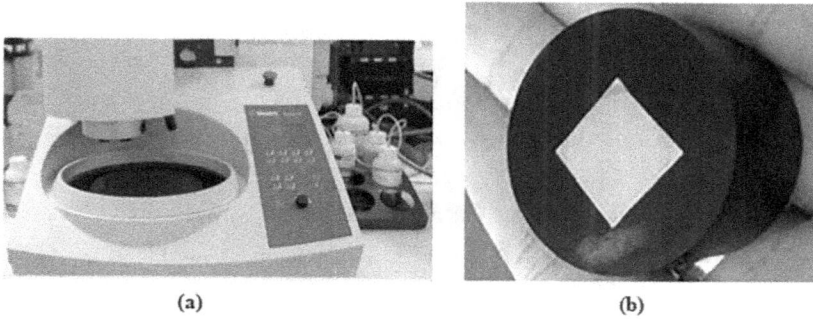

(a)   (b)

FIGURE 12.6   The automatic metallographic polishing machine (a), and polished metallographic specimen (b)

Solution

$L = 500$ mm $= 0.5$ m; $S = 15$ strokes/min; $V_a = $?
   By using Equation 12.1,

$$V_a = 2LS = 2 \times 0.5 \times 15 = 15 \, \text{m/min}$$

The honing tool's reciprocating speed $= V_a = 15$ m/min.

EXAMPLE 12.2: CALCULATING THE SURFACE SPEED OF THE ROTATING HONING TOOL

A honing operation is performed on a 120-mm bore cylinder by using a honing tool rotating at 90 rev/min. Calculate the surface/tangential speed of the honing tool.

Solution

$D = 120$ mm $= 0.12$ m, $N = 90$ rev/min, $V_r = $?
   By using Equation 12.2,

$$V_r = \pi\,D\,N = 3.142 \times 0.12 \times 90 = 34 \, \text{m/min}$$

The surface/tangential speed of the honing tool $= V_r = 34$ m/min

EXAMPLE 12.3: COMPUTING THE RESULTANT OF RECIPROCATING AND SURFACE SPEEDS OF HONING TOOL

By using the data in Examples 12.1–12.2, calculate the resultant of reciprocating and surface speeds of the honing tool.

Solution

The reciprocating speed $= V_a = 15$ m/min, and surface speed $= V_r = 34$ m/min, $V = $?

By using Equation 12.3,

$$V = \sqrt{V_a^2 + V_r^2} = \sqrt{15^2 + 34^2} = 37\,\text{m/min}$$

The resultant of reciprocating and surface speeds of the honing tool $= V = 37$ m/min

EXAMPLE 12.4: CALCULATING THE CROSSHATCH ANGLE IN A HONING OPERATION

By using the data in Examples 12.1–12.2, calculate the crosshatch angle in the operation.

Solution
$V_a = 15$ m/min, and $V_r = 34$ m/min, $\theta_h = ?$
 By using Equation 12.4,

$$\text{The crosshatch angle} = \theta_h = 2\tan^{-1}\left(\frac{V_a}{V_r}\right) = 2\tan^{-1}\left(\frac{15}{34}\right) = 2 \times 23.8° = 47.6°$$

EXAMPLE 12.5: CALCULATING THE ABRASIVE CONCENTRATION IN A LAPPING SLURRY

A lapping slurry is prepared by mixing 0.007 g of silicon carbide (SiC) abrasive with 0.051 g of kerosene oil (carrier fluid). Calculate the abrasive concentration in the slurry.

Solution
$m_a = 0.007$ g, $m_l = 0.051$ g, $C = ?$
 By using Equation 12.5,

$$C = \frac{m_a}{m_a + m_l} = \frac{0.007}{0.007 + 0.051} = 0.121$$

The abrasive concentration $= 0.121 = 12.1\%$

EXAMPLE 12.6: CALCULATING THE VOLUMES OF ABRASIVE MATERIAL AND CARRIER FLUID FOR LAPPING SLURRY

By using the data in Example 12.5, calculate the volumes of the following materials used to prepare the lapping slurry: (a) abrasive material and (b) carrier fluid. The density of silicon carbide is 3.2 g/cm³ and that of kerosene oil is 0.8 g/cm³.

Solution

$$\text{Density of the abrasive material} = \rho_a = 3.2\,\text{g/cm}^3, m_a = 0.007\,\text{g}, m_l = 0.051\,\text{g},$$
$$\rho_l = 0.8\,\text{g/cm}^3$$

Solution

a. Volume of the abrasive material $= V_a = \dfrac{m_a}{\rho_a} = \dfrac{0.007}{3.2} = 0.00218 \text{ cm}^3$

b. Volume of the carrier fluid $= V_l = \dfrac{m_l}{\rho_l} = \dfrac{0.051}{0.8} = 0.0637 \text{ cm}^3$

EXAMPLE 12.7: CALCULATING THE MRR (IN MM³/S) IN A LAPPING OPERATION

A force of 6 N is applied by the *lap* on the workpiece in a lapping operation. The relative speed of the lap and the work is 32 m/s. Calculate the MRR (in μm³/s) in the lapping operation, if the Preston's constant is 0.6×10^5 μm³/N-m.

Solution

$p = 6$ N, $v = 32$ m/s, $K = 0.6 \times 10^5$ (μm)³/N-m, MRR = ?
By using Equation 12.6,

$$\text{MRR}_{lapping} = K\,v\,p = 0.6 \times 10^5 \times 32 \times 6 = 1.15 \times 10^7 \left(\mu m\right)^3 / s$$

The MRR in the lapping operation $= 1.15 \times 10^7$ (μm)³/s

EXAMPLE 12.8: CALCULATING THE MRR IN LAPPING OPERATION

A lapping operation takes 4 minutes to finish a specified surface area of a metallic workpiece. Calculate the MRR if the height of material removed from the surface is 14 μm.

Solution

$h = 14$ μm, $T_m = 4$ min., MRR = ?
By using Equation 12.6,

$$\text{MRR}_{lapping} = \frac{h}{T_m} = \frac{14}{4} = 3.5\,\mu m/min$$

The MRR in the lapping operation $= 3.5$ μm/min

EXAMPLE 12.9: CALCULATING THE PERIPHERAL SPEED AND RESULTANT SPEED IN SUPERFINISHING

In a superfinishing operation, the 100-mm-diameter workpiece rotates at 200 rev/min. The stone reciprocates at 8 m/min. Calculate the resultant of the stone's reciprocating speed and the work's peripheral speed.

Solution

$D = 100$ mm, $N = 200$ rev/min, $V_a = 8$ m/min, $V = ?$
The peripheral speed of the work $= V_r = \pi\,D\,N = 3.142 \times 100 \text{ mm} \times 200 \text{ rev/min} = 62.84$ m/min

By reference to Figure 12.5,

$$\text{The resultant speed} = V = \sqrt{V_a^2 + V_r^2} = \sqrt{8^2 + 62.84^2} = 63.38\,\text{m/min}$$

EXAMPLE 12.10: CALCULATING THE MACHINING TIME IN A SUPERFINISHING OPERATION

A superfinishing operation is to be performed on a 500-mm-long workpiece rotating at 180 rev/min. The longitudinal feed of the stone is 3 mm/rev, and the depth of cut is 0.002 mm. The machining allowance is 0.011 mm. Calculate the machining time in the superfinishing operation.

Solution

$L = 500$ mm, $N = 180$ rev/min, $f = 3$ mm/rev, $d = 0.002$ mm, $a_m = 0.011$ mm, $T_{sf} = $?
By using Equation 12.7,

$$T_{sf} = \frac{L\,a_m}{f\,N\,d} = \frac{500 \times 0.011}{3 \times 180 \times 0.002} = 5.1\,\text{min}$$

The superfinishing time = 5.1 min.

QUESTIONS AND PROBLEMS

12.1   Tabulate the surface roughness values obtained by various abrasive finishing processes.

12.2   a. Diagrammatically illustrate the honing operation.
       b. List at least three components that are generally finished by honing.

12.3   a. Briefly explain the lapping operation with the aid of a sketch.
       b. Highlight the advantages and applications of lapping.

12.4   Define superfinishing operation with the aid of a sketch.

P12.5 A lapping slurry is prepared by mixing 0.008 g of silicon carbide (SiC) abrasive with 0.06 g of kerosene oil (carrier fluid). Calculate the abrasive concentration in the slurry.

P12.6 A honing operation is performed on a 120-mm bore cylinder by using a honing tool rotating at 90 rev/min. The stroke length of the honing tool is 500 mm, and the tool reciprocates 15 times during the operation. Calculate the (a) reciprocating speed of the tool, (b) surface/tangential speed of the tool, (c) resultant of reciprocating speed and surface speed, and (d) crosshatch angle in the operation.

P12.7 A force of 5 N is applied by the *lap* on the workpiece in a lapping operation. The relative speed of the lap and the work is 30 m/s. Calculate the MRR (in $\mu\text{m}^3/\text{s}$) in the lapping operation, if the Preston's constant is 0.6×10^5 $\mu\text{m}^3/\text{N-m}$.

P12.8 In a superfinishing operation, a 120-mm-diameter workpiece rotates at 190 rev/min. The stone reciprocates at 7 m/min. Calculate the resultant of the stone's reciprocating speed and the work's peripheral speed.

P12.9 A superfinishing operation is to be performed on a 480-mm-long workpiece rotating at 170 rev/min. The longitudinal feed of the stone is 2.5 mm/rev, and the depth of cut is 0.003 mm. The machining allowance is 0.011 mm. Calculate the superfinishing time.

MCQS

12.10 Underline the most appropriate answers for the following statements:
   a. Which abrasive finishing process involves the use of loose abrasive particles?
   (i) Honing, (ii) lapping, (iii) superfinishing, (iv) polishing
   b. Which abrasive finishing process is the most suitable for machining cylinder bores?
   (i) Honing, (ii) lapping, (iii) superfinishing, (iv) polishing
   c. Which abrasive finishing process involves the use of impregnated cloth?
   (i) Honing, (ii) lapping, (iii) superfinishing, (iv) polishing
   d. Which abrasive finishing process involves the use of a single abrasive stone?
   (i) Honing, (ii) lapping, (iii) superfinishing, (iv) polishing

REFERENCES

Deaconwscu, T., Deaconwscu, A. (2020), Developing an analytical model and computing tool for optimizing lapping operations of flat objects made of alloyed steels, *Materials*, 13(6): 1343–1358.

El-Hofy, H. A-G. (2018), *Fundamentals of Machining Processes - Conventional and Nonconventional Processes*, 3rd Edition, CRC Press, Boca Raton, FL, USA.

Edberg, S. and Landqvist, E. (2015), The impact of honing process parameters on the surface quality of cylinder liners, Master's Thesis, Department of Production Engineering, KTH Royal Institute of Technology, Stockholm, Sweden.

Huda, Z. (2020), *Metallurgy for Physicists and Engineers: Fundamentals, Analysis, and Calculations*, CRC Press, Boca Raton, FL, USA.

Preston, F.F. (1927), The theory and design of plate glass polishing machines; *Journal of Society of Glass Technology*, 11, 214–256.

Zhang, Y., Niu, J., Yang, Y., Gong, J. (2012), Study on the impact of the honing machine reciprocating reversing acceleration upon reticulate pattern trajectory, Proceedings of the International Conference on Mechanical engineering and Materials Science (MEMS 2012), Atlantis Press, Paris.

Zhu, G. (2011), *Materials Science and Engineering*, Trans Tech Publications, Switzerland.

Part IV

Advanced/Non-Traditional Machining

13 Computer Numerically Controlled Machining

13.1 WHAT IS COMPUTER NUMERICAL CONTROL MACHINING?

Computer numerical controlled (CNC) machining is an automated removal of material from a work-part. It is a technique of computer-aided manufacturing (CAM) by automating control of a machine tool by using software embedded in a microcomputer attached to the machine tool. A CNC system consists of three basic components: (a) work-part program, (b) microcomputer or machine control unit, and (c) processing equipment. The work-part program is a series of coded instructions usually written in *G-codes*; these instructions may refer to feed rate, tool position and motion, etc. The work-part program is stored in and executed by a microcomputer, called the machine control unit (MCU). The MCU is attached to the processing equipment (such as lathe, milling machine, grinder, etc.) that performs the machining operation to produce a machined part (see Figure 13.1).

13.2 ADVANTAGES AND LIMITATIONS OF CNC MACHINES

CNC machining is a cost-effective technique to increase efficiency and productivity in the workplace. CNC machines are increasingly being used in manufacturing/machining of intricate and complex parts for diversified industries, such as marine, aerospace, automotive, and medical sectors. There are several types of CNC machines, including CNC lathes, CNC milling machine, CNC drill press, CNC grinding machine, and the like (see Figure 13.2). A CNC machine is considered to provide more precision, complexity, and repeatability than is possible with manual machining; other benefits include greater accuracy, speed and flexibility, and capabilities such as contour machining. On the other hand, CNC machines have some limitations, which include more expensive than manual machine tools, require more maintenance than other production methods, and compel companies to hire a skilled CNC programmer.

The growing popularity of CNC machining is owing to the many advantages of CNC machines in comparison to traditional methods; these advantages include (a) consistency and quality, (b) productivity, (c) cost effectiveness, (d) safety, (e) versatility, (f) diversity in work material, and the like. These advantages are justified in the following paragraphs.

Consistency and Quality. A *CNC machine* is highly consistent and accurate in the work it produces since human error is almost completely eliminated. This makes CNC machining crucial for areas where quality is critical, since the machined part has a high level of reliability and quality.

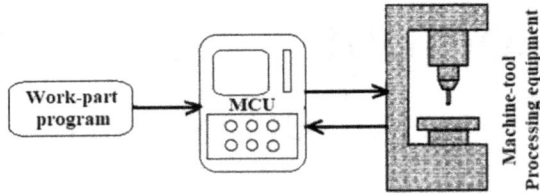

FIGURE 13.1   Basic components of a CNC machining system. (MCU = machine control unit)

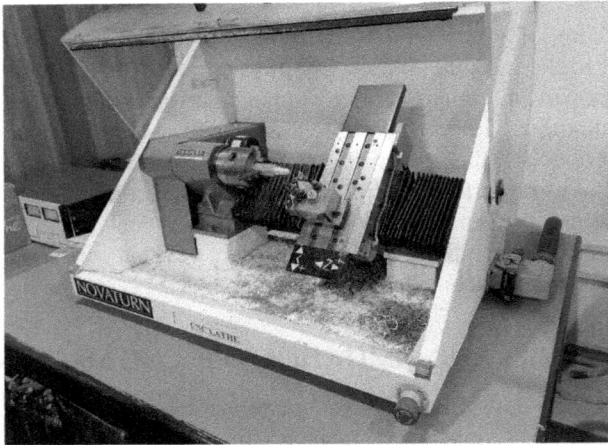

FIGURE 13.2   A CNC Lathe. (*Courtesy*: NED University of Engineering & Technology, Pakistan)

Productivity. It is possible to step away while the *CNC* machine tool gets to work since the tool can be programmed to perform a complex series of machining operations. Hence, manufacturing may be continued during out-of-hours automated machining in certain set-ups thereby increasing output rate and productivity. Thus several *CNC* centers can be simultaneously operated by engineers.

Cost Effectiveness. The return on investment time on CNC machines is short owing to a high rate of output and lower number of mistakes in resulting components. Additionally, machine operators require less training to operate a CNC machine and can learn how to use the machine in a virtual environment thereby eliminating the need for training (sample) workpieces. With growing demand and mass production of CNC machines, their cost will continue to drop.

Safety. CNC machines ensure safety to the operator, unlike conventional open-guard machining. In case of jam, machining error, or any other hazardous safety issue, the damage is caused to the machine itself and not to the operator.

Versatility. Thus CNC machines are versatile in use since they can be reprogrammed in a short period of time to produce a completely new product, making them ideal for short or long production runs. The part program, in CNC machining, can be changed in a short period of time without it being too costly or time-consuming.

Diversity in Work Material. CNC machine tools are capable of machining a wide range of workpiece materials, including metals, plastics, wood, foam, fiberglass, and the like.

13.3 COMPUTER-AIDED DESIGN (CAD) IN CNC MACHINING

13.3.1 What Is CAD?

Computer-aided design (CAD) involves the use of a computer for creating computer models defined by geometric parameters. These models are typically displayed on a computer monitor as a 3-dimensional (3D) representation of a work-part or a system of work-parts. A CAD software application (e.g. *TinkerCAD, SolidWorks, AutoCAD,* etc) enables the designer to analyze the model, modify it, and optimize it to create a detailed production drawing. CAD software packages enable designers to view products/machined parts under 2D and 3D representations and to test these products by CNC simulation (Fahad and Huda, 2017)

13.3.2 Role of CAD in CAM/CNC Machining

We learnt in Section 13.1 that CNC machining is a technique of computer-aided manufacturing (CAM) by automating control of a machine tool. In CNC machining, a G-coded *part program* is written and loaded onto the MCU of a CNC machine. Prior to *part programming,* engineers create a CAD drawing of the part to be manufactured/machined and then translate the drawing into *G-coded* program. The *part program* is loaded onto the *MCU* of the CNC machine, which interprets the drawing's coded instructions in CAD for machining the part. Then, a human operator performs a test run without the raw material in place (called *CNC simulation*) so to ensure proper positioning and performance. This step is important because incorrect speed or positioning can damage both the machine and the part. This author has experience of performing *CNC simulation* by running the *MTS-TOPCAM* software package at Nilai University, Malaysia. Today's CNC systems are integrated with CAD and CAM software, which can speed up the process of programming. The role of CAD in CAM/CNC machining is illustrated in Figure 13.3.

13.4 CNC MACHINE – *WORKING SYSTEM*

We learnt in Section 13.1 that there are three basic components of a CNC system: work-part program, MCU, and processing equipment. In this section, we will comprehend the working mechanism of the CNC system. Figure 13.4 illustrates the working mechanism of the CNC system with particular reference to turning machining operation.

It is evident in Figure 13.4 that the *work-part program* is stored in and executed by a microcomputer – *MCU*. The MCU is attached to the processing equipment (here, a CNC lathe) that performs the machining operation to produce a machined part. Now, the working mechanism(s) of each basic component is explained in the following sub-sections.

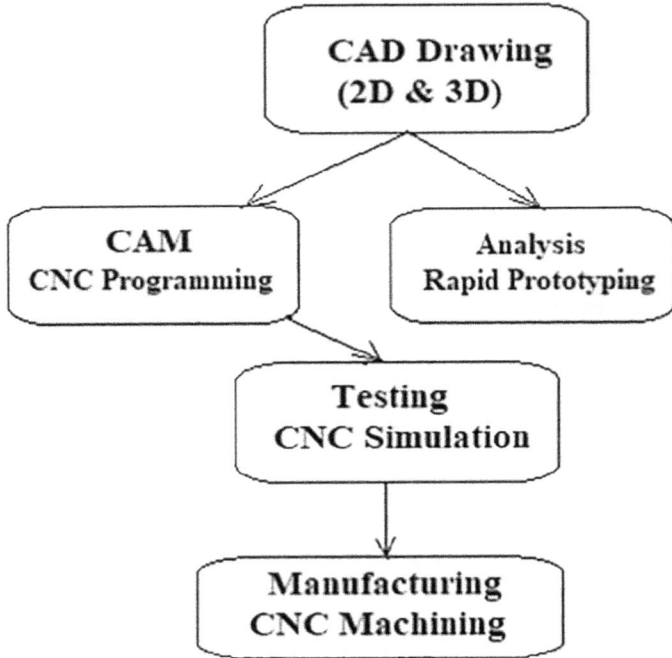

FIGURE 13.3   The role of CAD in CAM/CNC machining

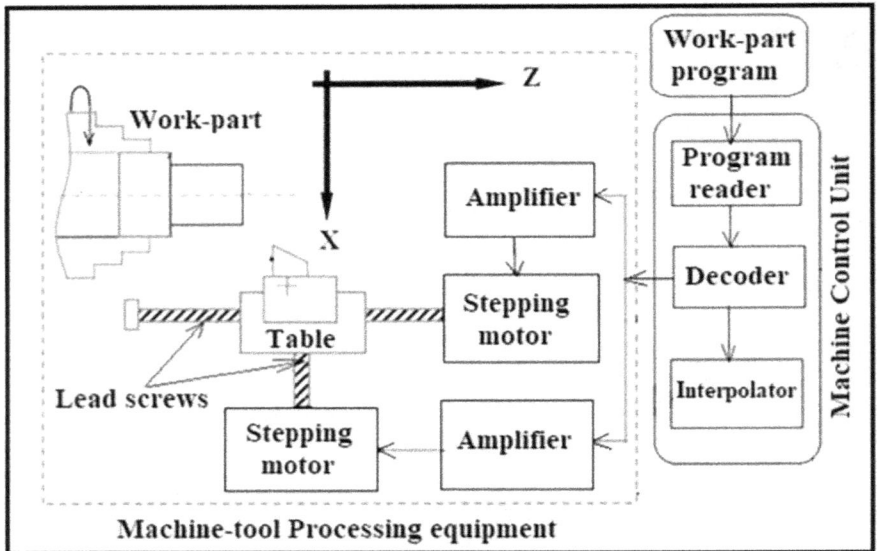

FIGURE 13.4 A CNC (open loop) system for turning machining operation. (MCU = machine control unit)

13.4.1 G-Codes and Work-Part Program

13.4.1.1 Code Words in a Part Program

The work-part program, in CNC programming, involves the use of code words that have been standardized as ISO 6983. In order to prepare a functional block or *syntax* for a work-part program, a combination of code words is generally used as follows: "N" word for the sequence number of a block; "G" word for the instructions related to movements of tool/workpiece; "T" word for selecting desired tool; "F" word for either feed (in mm/rev) or feed rate (in mm/min); "M" word for machine-related functions (*e.g.* "M07" for coolant ON); X, Y, and Z indicate linear axes, and A, B, and C represent rotational axes. Some commonly used G-codes are presented in Table 13.1.

The information in Table 13.1 can be used to write a part-program's block or syntax. For example, in case it is desired to compensate for a 14-mm-diameter tool to the left side of the work-part while moving the tool in a straight line at a feed rate of 100 mm/min to a position defined by X = 25 and Y = 20, the syntax will be as follows: G41 G01 X25 Y20 D14 F100 (see Example 13.1). The use of G-codes is further illustrated in the following sub-section. Besides G-codes, M-codes (machine codes) are also important in CNC programming. The commonly used M-codes are listed in Table 13.2.

13.4.1.2 Work-Part Program in CNC Programming

A *work-part program* is a set of commands (coded instructions) to be followed by the processing equipment to produce a work-part (Smid, 2007). Each command, in the program, specifies a position or motion that is to be accomplished by the tool relative to the work. The G-coded instructions in the work-part program comprise of letters, numbers, and symbols that are arranged in a format of functional blocks (Manton and Weidinger, 2010). There are numerical data associated with each functional block for

TABLE 13.1
Some Commonly used G-codes and Their Meanings

G-code	Meaning
G00	Positioning (rapid traverse)
G01	Linear interpolation
G02	Circular interpolation: clockwise (CW) arc
G03	Circular interpolation: counter-clockwise (CCW) arc
G17, G18, and G19	Plane designations (XY, ZX, and YZ)
G21	Metric units
G27	Return to reference point
G28	Tool at original position
G40	To cancel a cutter diameter compensation
G41	Compensation of cutter diameter to the left
G42	Compensation of cutter diameter to the right
G54	Work offset
G70	Dimensions in inches
G90	Absolute positioning
G91	Incremental positioning
G94	Feed rate units: mm/min
G95	Feed: mm/rev

TABLE 13.2

The Commonly used M-codes for CNC Programming

M-code	Meaning
M00	Program stop
M01	Optional stop
M02	End of program
M03	Spindle start clockwise (CW)
M04	Spindle start CCW
M05	Spindle stop
M06	Tool change
M07	Mist coolant ON
M08	Flood coolant ON
M09	Coolant OFF
M10	Clamp
M11	Unclamp
M13	Spindle CW Coolant ON
M14	Spindle CCW Coolant ON

processing a segment of the workpiece; these numerical data refers to the machining parameters, including spindle rpm, cutting speed (in m/min), spindle direction, feed (mm/rev), feed rate (mm/min), tool change instructions, and other machine-related commands (see Example 13.2). Sometimes it is required to specify spindle rotational speed in rpm; this is represented as follows: G97 S1800 (the spindle rotational speed = 1800 rpm). In case a constant cutting speed of 160 m/min is desired, the representation is as follows: G96 S160 (see Example 13.3). The significance of G-codes (Table 13.1) and M-codes (Table 13.2) is further illustrated in the last example in Section 13.9.

13.4.2 MACHINE CONTROL UNIT

The machine control unit (MCU) consists of three main components: (a) program reader, (b) decoder, and (c) interpolator (see Figure 13.4). Once *work-part program* has been loaded onto *MCU*, the *program reader* reads the coded instructions in the program, interprets it, and conveys it to *decoder*. The *decoder* decodes the instruction and generates pulses that are required to drive the stepping motors (in case of open-loop CNC system). Sometimes, simultaneous travel of the tool post in two directions is required; this objective is achieved by *interpolator* which implements simultaneous travel of the tool post (*e.g.* in the *X* and *Z* directions) to achieve linear, circular, and helical motions of the tool.

Besides the basic functions mentioned in the preceding paragraph, there are some auxiliary functions of MCU. The MCU sends the control signals (pulses), which represent the position and speed of each axis, to the amplifier circuits (in the processing equipment). It also implements auxiliary control functions, such as spindle ON/OFF, coolant ON/OFF, tool change, and the like. In case of a closed-loop CNC system, the MCU also receives the feedback signals of position and speed for each drive axis. It also allows the programmer to edit the part program in the event that the program contains errors or changes in cutting conditions are required.

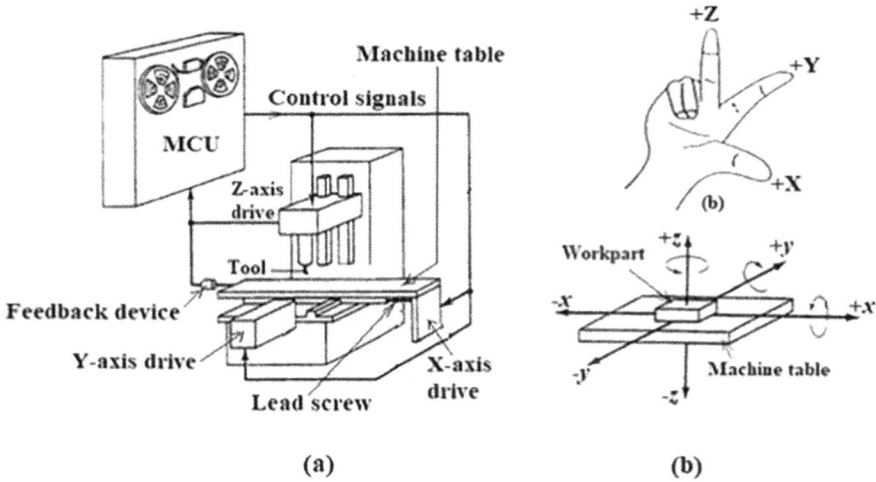

FIGURE 13.5   A CNC milling machine's working mechanism (a), right-hand coordinate rule (b), and the three linear axes: X, Y, Z for the CNC machine's worktable (c)

13.4.3 Processing Equipment in CNC Machining

The processing equipment is composed of the drive system (amplifiers, stepping motor or servomotor, and ball lead screws) and the machine tool (machine-table, cutting tool, work holding device, workpiece, etc.) that accomplish the sequence of processing steps (machining operations) to transform the starting work-part into finished part. It is evident in Figure 13.4 that the amplifiers receive the control signals (pulses) that are sent by MCU. These pulses are amplified to actuate stepping motors resulting in the rotation of the motor-spindle, which in turn rotate the lead screws to position the machine-table (see Figure 13.4). Each pulse, sent from the MCU, causes the lead screw to rotate a fraction of one revolution – called the *step angle*. The number of pulses transmitted to each axis is equal to the required amount of motion. One revolution of the lead screw results in one pitch (mm) linear motion of the machine-table (Overby, 2010).

A right-hand coordinate system is used to describe the motions along the three linear axes (X, Y, Z) for a CNC machine tool. The three linear axes (X,Y, Z) and the working mechanism of a CNC milling machine is illustrated in Figure 13.5.

13.5   ZERO SYSTEMS AND POSITIONING SYSTEMS IN CNC MACHINING

13.5.1 Zero Systems in CNC Machining

A CNC machine has at least one *fixed reference point*, also called *fixed machine zero* (0,0,0). The *fixed machine zero* is a datum point characteristic of the machine tool at which the X, Y and Z axes of the machine tool intersect. When it comes to deciding reference point or zero point for the work-part or the cutting tool, the CNC

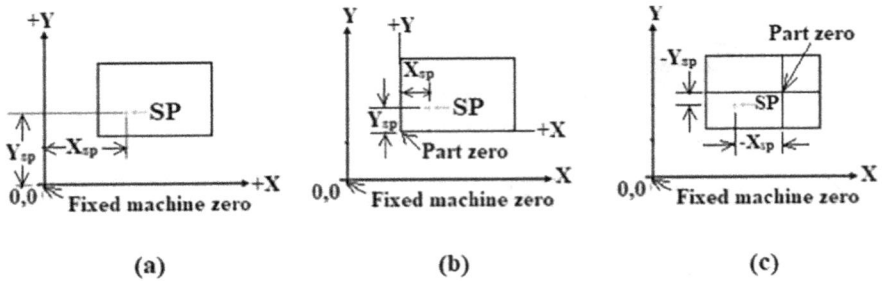

FIGURE 13.6   The fixed zero system (a), full zero shift system (b), and full floating zero system (c)

programmer has a certain degree of freedom. A *part reference point* or *program zero* is a datum point characteristic of the work-part; it must be located accurately with respect to the *fixed machine zero* to define the part position in the machine coordinate system. A *program zero* is a flexible point *i.e.* its actual position is in the programmer's hands. There are three commonly used zero systems: (a) fixed zero system, (b) full zero shift system, and (c) full floating zero system. In the *fixed zero system*, the *part zero* or *program zero* is at the *fixed machine zero* point (Figure 13.6a). In the full zero shift system, the operator clamps the work-part at a convenient position in the worktable (Figure 13.6b). In the *full floating zero system*, the *part zero* can be moved to any desirable point (Figure 13.6c).

13.5.2 COORDINATE POSITIONING SYSTEMS

There are two types of CNC positioning systems: (a) absolute positioning system and (b) incremental positioning system. In *absolute positioning system*, all the motion commands are referred from a reference point or origin of the axis system. In the *incremental positioning system*, the distance of each succeeding location is measured from the previous location. For example, if we want to drill two holes at two locations with coordinates $A(x_1,y_1)$ and $B(x_2,y_2)$. The motion specification for absolute positioning will be:

$$\text{Absolute positioning}: X = x_2 \text{ and } Y = y_2 \tag{13.1}$$

In case of incremental positioning the move is specified by:

$$\text{Incremental positioning}: X = x_2 - x_1 \text{ and } Y = y_2 - y_1 \tag{13.2}$$

The significance of Equations 13.1–13.2 is illustrated in Example 13.4.

13.5.3 POSITIONING SYSTEM

In the point-to-point positioning system, the tool or the work-part is moved from one point to another point, and then the tool performs the required task. Upon completion of the one machining operation, the tool (or work-part) moves to the next position and repeats the operation cycle. The simplest example for this type of system is a drilling operation, where the worktable moves. For example, in a drilling operation, if the worktable moves from position (x_1, x_2) to (y_1, y_2), the travel times in X-axis and Y-axis can be computed by:

$$t_x = \frac{x_2 - x_1}{v_x} \tag{13.3}$$

$$t_y = \frac{y_2 - y_1}{v_y} \tag{13.4}$$

where t_x and t_y are the travel times in X-axis and Y-axis, respectively, and v_x and v_y are the velocities of the worktable in the x-direction and y-direction, respectively (see Example 13.5).

13.6 CONTROL SYSTEMS IN CNC MACHINES

There are two types of control systems in CNC machines: (a) open-loop control CNC system and (b) closed-loop control CNC system. Figure 13.7 illustrates the two types of control CNC systems.

13.6.1 OPEN-LOOP CONTROL CNC SYSTEM

The open-loop control system is the control system with no feedback mechanism *i.e.* an *open-loop control CNC system* operates without verifying that the desired position of the worktable has been achieved. The open-loop control CNC system uses stepping motors for driving the lead screw. A stepping motor is a device whose output shaft rotates through a fixed angle in response to an input pulse (see Figure 13.7a). The open-loop system lacks accuracy since its accuracy depends on the motor's ability to step through the exact number. This system is more suitable for cases where a significant tool force does not exist (*e.g.* laser cutting).

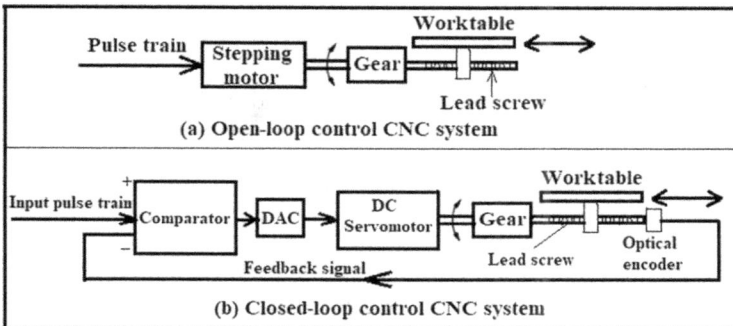

FIGURE 13.7 Loop control systems in CNC machines. (DAC = digital-to-analog convertor)

13.6.2 Closed-Loop Control CNC System

A closed-loop control CNC system operates by verifying that the desired position of the worktable has been achieved. In this systems (Figure 13.7b), the DC servomotors and feedback devices are used to ensure that the desired position is achieved. The feedback sensor used is an optical encoder, which consists of a light source, a photo-detector, and a disk containing a series of slots. The encoder is connected to the lead screw. As the lead screw turns, axis-position feedback signals are issued by the optical encoder (or position transducer) for error correction to the *comparator*; which compares the actual results with the desired results and issues corrective signals to the worktable via a digital-to-analog converter (DAC), amplifier, and servomotor. The closed-loop CNC systems have a high degree of accuracy. They are appropriate when there is a force resisting the movement of the tool/work-part, such as in CNC milling machines, CNC lathes, and the like. Recently (2019), an advanced closed-loop CNC system has been developed that enables measurement of work-part location and geometry before and after machining and therefore ensures work-part geometry traceability (Sulak and Bracun, 2019).

13.7 MATHEMATICAL MODELING OF OPEN-LOOP CONTROL CNC MACHINING

The MCU unit issues *pulses* (pulse train) to implement the required motion in the processing equipment's motor. *Each pulse* causes the motor's shaft to rotate a *fraction of one revolution*, called the *step angle*. The step angle α (in deg/step) can be determined by:

$$\alpha = \frac{360°}{n_s} \tag{13.5}$$

where n_s is the number of step angles for the motor.

The gear ratio (r_g) is the ratio of the rotational speed of the motor's output shaft (N_m) to the rotational speed of the lead screw (N_{ls}). Mathematically, the gear ratio can be expressed by:

$$r_g = \frac{N_m}{N_{ls}} = \frac{A_m}{A_{ls}} \tag{13.6}$$

where A_m is angle of motor-shaft rotation, deg.; and A_{ls} is the angle of lead screw rotation, deg. The angle of motor-shaft rotation (A_m) can be computed by:

$$A_m = n_p \alpha \tag{13.7}$$

where α is the step angle, deg./step, and n_p is the number of pulses received by the step motor.

The distance moved by the worktable in x-direction can be calculated by:

$$x = \frac{p \cdot A_b}{360} \tag{13.8}$$

where x is the distance moved by the worktable in the x-direction, mm; p is the pitch of the ball screw, mm/rev; and A_{ls} is the angle of lead screw rotation, deg.

The number of pulses received by the motor (n_p) can be related to the gear ratio (r_g) as follows:

$$n_p = \frac{360 \cdot x \cdot r_g}{p \cdot \alpha} \tag{13.9}$$

The significance to Equation 13.9 is illustrated in Example 13.6. Sometimes, the lead screw is directly connected to the output shaft of the stepping motor; in this case, the gear ratio is 1:1 so Equation 13.9 reduces to:

$$n_p = \frac{360 \cdot x}{p \cdot \alpha} \tag{13.10}$$

The significance of Equation 13.10 is illustrated in Example 13.7.

The table travel speed or the feed rate f_r (in mm/min) can be calculated by:

$$f_r = N_{ls} \cdot p \tag{13.11}$$

where N_{ls} is the rotational speed of the lead screw; and p is the lead screw pitch, mm/rev.

The pulse train frequency f_p (*in Hz*) can be calculated by:

$$f_p = \frac{\left(f_r \cdot n_s \cdot r_g \right)}{\left(60 \cdot p \right)} \tag{13.12}$$

where n_s is the number of step angles of the motor and r_g is gear ratio (see Example 13.9).

13.8 MATHEMATICAL MODELING OF CLOSED-LOOP CNC SYSTEM

It is learnt in Section 3.6 that the closed-loop control CNC system has an optical encoder to sense errors and issue feedback signals (see Figure 13.7b). The number of slots in the encoder/rev (= the number of pulses issued/revolution of the lead screw) n_s is computed by:

$$n_s = \frac{360°}{\alpha} \tag{13.13}$$

where α is angle between slots in the optical encoder, deg/pulse.

The number of pulses sensed by comparator (n_p) is related to angle between slots α by:

$$n_p = \frac{A_{ls}}{\alpha} \tag{13.14}$$

By combining Equations 13.13 and 13.14, we obtain:

$$n_p = \frac{\left(A_{ls} \cdot n_s\right)}{360°} \tag{13.15}$$

where A_{ls} is the angle of lead screw rotation, deg.

The distance moved by the worktable in the x-direction (x, mm) can be calculated by:

$$x = \frac{\left(p \cdot n_p\right)}{n_s} \tag{13.16}$$

where n_p is the number of pulses sensed by count comparator; n_s is the number of pulses per revolution of the lead screw (see Example 13.10).

The table travel speed or feed rate and pulse train frequency can be calculated by using Equations 13.11 and 13.12, respectively.

13.9 CALCULATIONS – WORKED EXAMPLES ON CNC MACHINING

EXAMPLE 13.1: WRITING SYNTAX FOR CANCELLING A CUTTER-DIAMETER-COMPENSATION

In a drilling operation, it is desired to cancel for a 12-mm-diameter tool compensation by moving the cutting tool in a straight line at a feed rate of 110 mm/min to a position defined by X = 18 and Y = 22. Write the syntax (line of coded instructions) for the work-part program.

Solution
By reference to Table 13.1, the syntax for the work-part program is as follows:
N30   G90   G94
N40   G40   G01   X18   Y22   D12   F110.

EXAMPLE 13.2: WRITING SYNTAX FOR LINEAR INTERPOLATION FOR A WORK-PART PROGRAM

In a turning operation, the cutting tool (No. 2) must move in a straight line to a position with x-coordinate 5.0 mm and y-coordinate 7.0 mm. The tool's feed is 4.5 mm/rev. The mist coolant is ON. Write two syntax statements using absolute positioning system for the operation.

Solution

By reference to Tables 13.1 and 13.2, the syntax statements are:
N10   G90   G95
N20   G01   X5.0   Y7.0   T02   F4.5   M07

EXAMPLE 13.3: WRITING SYNTAX WITH CIRCULAR INTERPOLATION FOR A WORK-PART PROGRAM

In a turning operation, the cutting tool (No. 1) must move along a CCW arc to a position with x-coordinate of 3.5 mm and y-coordinate of 6.0 mm. The arc radius is 8 mm, and the tool feed rate is 110 mm/min. The rotational speed of spindle is 2000 rpm. Write two syntax statements.

Solution

By reference to Table 13.1, the two syntax statements are given below:
N10   G90   G94   G97
N20   G03   X3.5   Y6.0   R8.0   T01   S2000   F110

EXAMPLE 13.4: MOTIONS DEFINITIONS IN ABSOLUTE POSITIONING AND INCREMENTAL POSITIONING

Two holes are required to be drilled at positions $P_1(4,6)$ and $P_2(12,16)$ in a metal plate.

(a) Define the motion of the work-part in (i) absolute positioning system and (ii) incremental positioning system. (b) Graphically show the motion defined in (a) above.

Solution

$x_1 = 4, y_1 = 6, x_2 = 12, y_2 = 16, X = ?, Y = ?$

a. i. By using the set of Equations 13.1,

$$\text{Absolute positioning:}\quad X = x_2 = 12, Y = y_2 = 16$$

ii. By using the set of Equations 13.2,

$$\text{Incremental positioning:}\quad X = x_2 - x_1 = 12 - 4 = 8 \text{ and } Y = y_2 - y_1 = 16 - 6 = 10$$

b. The graphical representation of the motion is shown in Figure E-13.4.

EXAMPLE 13.5: CALCULATING THE TRAVEL TIME FOR POINT TO POINT IN CNC DRILLING MACHINE

The XY worktable of a drilling machine has to be moved from the point (2,1) to (17,20). The velocity of worktable along each axis is 10 mm/s. Calculate the travel time for the motion.

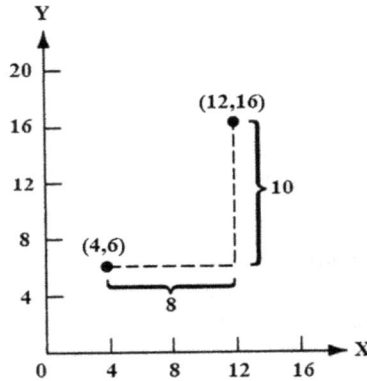

FIGURE E-13.4 The graphical representation of absolute positioning and incremental positioning

Solution

$x_1 = 2$, $y_1 = 1$, $x_2 = 17$, $y_2 = 20$, $v_x = v_y = 10$ mm/s, travel time $= ?$

By using Equation 13.3 and Equation 13.4,

$$t_x = \frac{x_2 - x_1}{v_x} = \frac{17 - 2}{10} = 1.5\,\text{s}$$

$$t_y = \frac{y_2 - y_1}{v_y} = \frac{20 - 1}{10} = 1.9\,\text{s}$$

The travel time $= 1.9\,\text{s} \cong 2\,\text{s}$

EXAMPLE 13.6: COMPUTING THE NUMBER OF PULSES REQUIRED WHEN THE LEAD SCREW IS CONNECTED TO MOTOR SHAFT THROUGH A GEAR-BOX

The worktable of an open-loop CNC system is driven by a lead screw that has 4 mm pitch. The lead screw is connected to the output shaft of a stepping motor, having 45 step angles, through a gear-box with gear ratio of 4:1. It is required to move the work-table a distance of 180 mm from its present position at a table travel speed of 230 mm/min. Calculate the number of pulses required.

Solution

$p = 4$ mm, $n_s = 45$, $r_g = 4$, $x = 180$ mm, $n_p = ?$

By using Equation 13.5,

$$\alpha = \frac{360°}{n_s} = \frac{360°}{45} = 8°$$

By using Equation 13.9,

$$n_p = \frac{360 \cdot x \cdot r_g}{p \cdot \alpha} = \frac{360 \times 180 \times 4}{4 \times 8} = 810$$

The required number of pulses $= n_p = 8100$

EXAMPLE 13.7: COMPUTING THE NUMBER OF PULSES REQUIRED WHEN THE LEAD SCREW IS DIRECTLY CONNECTED TO THE MOTOR SHAFT

Repeat Example 13.6 if the lead screw is directly connected to the motor shaft.

Solution
$p = 4$ mm, $x = 180$ mm, $\alpha = 8°$, $r_g = 1$, $n_p = ?$
By using Equation 13.10,

$$n_p = \frac{360 \cdot x}{p \cdot \alpha} = \frac{360 \times 180}{4 \times 8} = 2025 \text{ pulses}$$

EXAMPLE 13.8: CALCULATING THE STEPPING MOTOR RPM IN AN OPEN-LOOP CNC SYSTEM

By using the data in Example 13.6, calculate the required rotational speed of the motor.

Solution
$r_g = 4$, $n_p = 8100$, $f_r = 230$ mm/min, $p = 4$ mm, $N_m = ?$
By re-arranging the terms in Equation (13.11),

$$N_{ls} = \frac{f_r}{p} = \frac{230}{4} = 57.5 \text{ rev/min}$$

By re-arranging the terms in Equation (13.6),

$$N_m = r_g \cdot N_{ls} = 4 \times 57.5 = 230 \text{ rev/min}$$

The required rotational speed of the motor $= 230$ rpm.

EXAMPLE 13.9: CALCULATING THE PULSE TRAIN FREQUENCY IN AN OPEN-LOOP CNC SYSTEM

By using the data in Example 13.6, calculate the pulse train frequency corresponding to the desired table speed.

Solution
$r_g = 4$, $n_p = 8100$, $f_r = 230$ mm/min, $p = 4$ mm, $n_s = 45$
By using Equation (13.12),

$$f_p = \frac{\left(f_r \cdot n_s \cdot r_g \right)}{\left(60 \cdot p \right)} = \frac{230 \times 45 \times 4}{60 \times 4} = 172.5 \text{ Hz}$$

EXAMPLE 13.10: CALCULATING THE NUMBER OF PULSES SENSED BY THE COUNT COMPARATOR

The lead screw of a closed-loop CNC system has a pitch of 4.5 mm and is coupled to a servomotor with a gear ratio of 4:1. The MCU of the CNC machine issues 90 pulses per revolution of the lead screw. The worktable is programmed to move a distance of 80 mm at a speed of 430 mm/min. Calculate the number of pulses sensed by the count comparator to verify the exact 80-mm movement of the worktable.

Solution

$p = 4.5$ mm, $r_g = 4$, $n_s = 90$ pulses/rev., $x = 80$ mm, $f_r = 430$ mm/min, $n_p = ?$
By re-arranging the terms in Equation (13.16),

$$n_p = \frac{(x \cdot n_s)}{p} = \frac{(80 \times 90)}{4.5} = 1600 \text{ pulses}$$

The number of pulses sensed by the count comparator $= n_p = 1600$ pulses

EXAMPLE 13.11: CALCULATING THE PULSE TRAIN FREQUENCY IN A CLOSED-LOOP CNC SYSTEM

By using the data in Example 13.10, calculate the pulse train frequency.

Solution

$p = 4.5$ mm, $r_g = 4$, $n_s = 90$ pulses/rev., $x = 80$ mm, $f_r = 430$ mm/min, $f_p = ?$
By using Equation (13.12),

$$\text{Pulse train frequency} = f_p = \frac{(f_r \cdot n_s \cdot r_g)}{(60 \cdot p)} = \frac{(430 \times 90 \times 4)}{(60 \times 4.5)} = 573 \text{ Hz}$$

EXAMPLE 13.12: DETERMINING COORDINATES OF MACHINING POINTS FOR PART PROGRAMMING

Figure E-13.12 shows a billet with dimensions ($300 \times 300 \times 25$ mm³). It is required to machine (by pocket milling) a section 250×250 square for a depth of 8 mm. (a) Assume work offset and the tool length compensation, (b) determine the coordinates of the corner points: I, J, K, and L.

Solution

In Figure E-13.12, G54 shows the tool's original position with the coordinate point (x,y): $(0,0)$.

   a. The assumptions are as follows: Work offset: G54, Tool length compensation: H01.

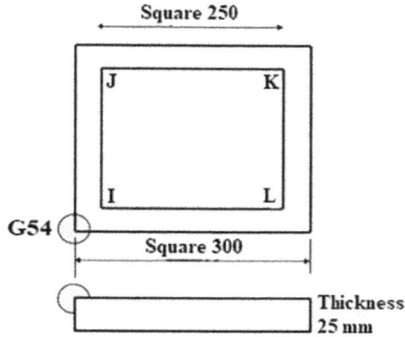

FIGURE E-13.12 Drawing of the $(300 \times 300 \times 25 \text{ mm}^3)$ billet for CNC machining

b. The x-coordinate of $I = (300 - 250)/2 = 25$; The y-coordinate of $I = (300 - 250)/2 = 25$.

Point I: X = 25 and Y = 25;   Point J: X = 25 and Y = 275
Point K: X = 275 and Y =275;   Point L: X = 275 and Y = 25

Example 13.13: Writing a Work-Part Program for CNC Machining

By using the data and the drawing in Example 13.12, write a part program for CNC machining of the billet section.

Solution
By reference to Tables 13.1 and 13.2, the *Part Program* for CNC machining is as follows:

P90	Program number
N10 G21 G94	Metric units input, Feed rate in mm/min
N20 G91 G28 X0 Y0 Z0	The tool is at original position
N30 T01 M06	The tool is 17-mm diameter End Drill
N40 G90 G54 G00 X25.0 Y25.0	Work offset call
N50 G43 H01	Tool length compensation call
N60 M03 S800 Z25.0	Position the tool above the starting point: *I*. The spindle rotates at 800 rpm.
N70 G01 Z−8.0 F350	The tool is linearly moved inside the work to a depth of 8 mm.
N80 G01 X25.0 Y275.0	The tool is moved to the corner point *J*.
N90 G01 X275.0	The tool is moved to the corner point *K*.
N100 G01 Y25.0	The tool is moved to the corner point *L*.
N110 G01 X25.0	The tool is moved to the corner point *I*.
N120 G00 Z100.0 M05	The tool is taken out of work; spindle stops.
N130 G91 G28 X0 Y0 Z0	Tool is moved back to the original position.
N140 M30	End of program

QUESTIONS AND PROBLEMS

13.1 Define a CNC machining system with the aid of a diagram.

13.2 a. List the advantages of CNC machines and explain one of them.
      b. What are the limitations of CNC machines?

13.3 Briefly explain the role CAD in CNC machining with the aid of a sketch.

13.4 Diagrammatically illustrate the working system of CNC machining for turning operation.

13.5 Explain the components and functions of machine control unit (MCU) in CNC machining.

13.6 a. List the components of the drive system and machine tool in the processing equipment
      b. State the functions of the components listed by you in (a) above.
      c. Draw a sketch showing the linear axes: X, Y, Z for a CNC milling machine's worktable.

13.7 Differentiate between an open-loop control and closed-loop control CNC systems with the aid of diagrams.

13.8 Explain the zero systems in CNC machining with the aid of sketches.

P13.9 In a turning operation, the cutting tool (No. 2) must move along a CW arc to a position with x-coordinate of 5.5 mm and y-coordinate of 4.0 mm. The arc radius is 10 mm, and the tool feed is 5 mm/rev. The work's constant cutting speed is 250 m/min. The mist coolant is ON. Write two syntax statements for the operation.

P13.10 The coordinate of points P_1 and P_2 in the absolute positioning system are $P_1(4,2)$ and $P_2(6,-2)$. What are the coordinates of P2 in the incremental positioning system?

P13.11 The XY worktable of a drilling machine has to be moved from the point (3,3) to (9,7). The velocity of worktable along each axis is 8 mm/sec. Calculate the travel time for the motion.

P13.12 The worktable of a CNC system is driven by a lead screw that has 3.5 mm pitch. The lead screw is connected to the output shaft of a stepping motor with 45 step angles. It is required to move the worktable to a distance of 160 mm from its present position at a table travel speed of 200 mm/min. Calculate the number of pulses required when (a) the lead screw is directly connected to the motor's output shaft and (b) the lead screw is coupled with a gear-box (gear ratio: 3:1).

P13.13 The lead screw of a closed-loop CNC system has a pitch of 4 mm and is coupled to a servomotor with a gear ratio of 4:1. The MCU of the CNC machine issues 100 pulses per revolution of the lead screw. The worktable is programmed to move a distance of 70 mm at a speed of 380 mm/min. Calculate the number of pulses sensed by the count comparator to verify the exact 70-mm movement of the worktable.

P13.14 The worktable of an open-loop CNC system is driven by a lead screw that has 5 mm pitch. The lead screw is connected to the output shaft of a stepping motor, having 47 step angles, through a gear-box with gear ratio of 4:1. It is required to move the worktable a distance of 170 mm from its present

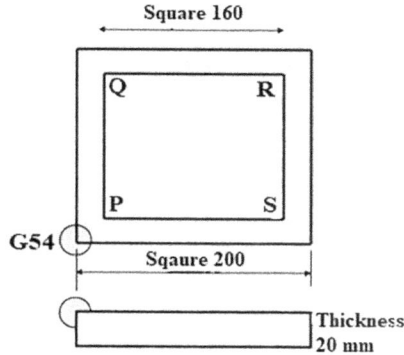

FIGURE P13.15   Sketch/Drawing of (200 mm × 200 mm × 20 mm) billet for CNC machining

position at table travel speed of 220 mm/min. Calculate the (a) required rotational speed of the motor, (b) the pulse train frequency corresponding to the desired table speed.

P13.15   Figure P13.15 shows a billet with dimensions (200 mm × 200 mm × 20 mm). It is required to machine (by pocket milling) a section 160 × 160 square for a depth of 6 mm. (a) Assume work offset and the tool length compensation and (b) determine the coordinates of the corner points: P, Q, R, and S.

P13.16   By using the data and the drawing in P13.15, write a part program for CNC machining of the billet section.

MCQS

P13.17a. Which component of the CNC system senses errors and issues feedback signals?
(i) Optical encoder, (ii) comparator, (iii) servomotor, (iv) lead screw
b. Which one is the essential component in both open-loop and closed-loop systems?
(i) Optical encoder, (ii) comparator, (iii) servomotor, (iv) lead screw
c. Which component issues corrective signals to digital-to-analog converter?
(i) Optical encoder, (ii) comparator, (iii) servomotor, (iv) lead screw
d. Which component receives corrective signals in a closed-loop control CNC system?
(i) Optical encoder, (ii) comparator, (iii) servomotor, (iv) lead screw

REFERENCES

Fahad, M., Huda, Z. (2017), Computer integrated manufacturing, In: (ed: Z. Huda) *Materials Processing for Engineering Manufacture*, Trans Tech Publications, Pfaffikon, Switzerland.

Manton, M. and Weidinger, D. (2010), *Computer Numerical Control Workbook – Generic Lathe*, CamInstructor, Kitchener, Ontario, Canada.

Overby, A. (2010), *CNC Machining Handbook*, McGraw Hill Education TAB, New York.

Smid, P. (2007), *CNC Programming Handbook*, 3rd Edition, Industrial Press, South Norwalk, CT, USA.

Sulak, L., Bracun, D. (2019), Evaluation of localization systems for *CNC machining* of large FRPC parts, *Procedia CIRP*, 812019, 844-849.

14 Mechanical Energy-based Machining Processes

14.1 WHAT ARE THE MECHANICAL ENERGY-BASED MACHINING PROCESSES?

We learnt in Chapter 1 that nontraditional machining (NTM) processes are chip-less material removal processes that involve the use of energy for material cutting. The NTM processes may be classified into four groups: (1) mechanical energy-based machining processes, (2) thermal energy-based machining processes, (3) electro-chemical machining processes, and (4) chemical machining process (Grzesik, 2016). The mechanical energy-based machining involves the removal of material by mechanical erosion or abrasion; these processes include (a) ultrasonic machining, (b) abrasive jet machining, (c) water jet machining, and (d) abrasive water jet machining. The mechanical energy-based NTM processes are explained in the following sections.

14.2 ULTRASONIC MACHINING (USM)

14.2.1 THE USM PROCESS AND ITS ADVANTAGES

USM Process. *Ultrasonic machining (USM)*, also called *ultrasonic vibration machining*, is a mechanical-energy-based non-traditional material removal process that involves the erosion of holes or cavities on a hard or brittle workpiece by using shaped tools, high-frequency mechanical vibration, and an abrasive slurry. In USM, a low-frequency electrical signal is applied to a transducer which converts the electrical energy into high-frequency mechanical vibration with a frequency in the range of 20–40 kHz. This high-frequency mechanical energy is then transmitted to a horn-and-tool assembly, which results in a unidirectional vibration of the tool at the ultrasonic frequency with a known amplitude in the range of 15–50 μm (see Figure 14.1a). A constant stream of abrasive slurry is also passed between the tool and workpiece. The abrasive hard particles in slurry are accelerated toward the surface of the workpiece by a tool oscillating at a frequency up to 40 kHz through repeated abrasions; the tool machines a cavity of a cross-section identical to its own (Figure 14.1b).

Advantages of and Limitations of USM Process. The *USM* process is capable of machining hard and brittle materials (both conductive and dielectric), such as single crystals, glasses, polycrystalline ceramics (*e.g.* boron carbide, titanium carbides, rubies, quartz, etc), and the like. The USM offers solution to the problems arising from increasing complex operations to provide intricate shapes and workpiece profiles. It is therefore used extensively in machining hard and brittle materials that are difficult to machine by traditional manufacturing processes. USM can

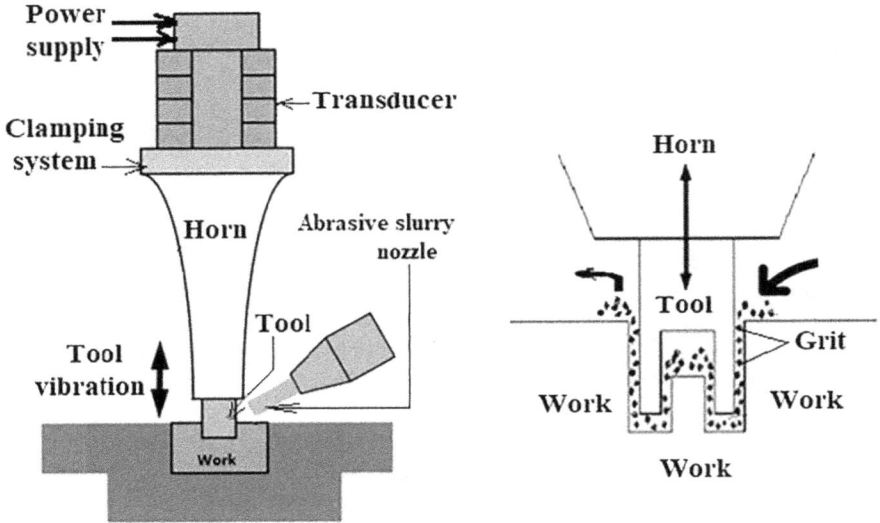

FIGURE 14.1 Schematic of ultrasonic machining process, (a) slurry motion and material removal (b)

also produce non-round holes with very small radius of curvatures. The parts, machined by USM, possess better surface finish and higher structural integrity. There is no or little thermal, electrical or chemical damage on the surface machined by USM (Kuo et al., 2013). However, there are some limitations of USM. For example, USM has higher power consumption and lower material removal rates (MRRs) than traditional machining processes, and the tool wears fast. Sometimes there are edge chipping issues in USM machined parts, but this problem can be resolved by using electrophoresis-assisted micro-ultrasonic machining (EPAMUSM) (He et al., 2018).

14.2.2 MATHEMATICAL MODELING OF USM

In USM, there are two principal mechanisms of material removal due to grit (abrasive particle) action: (a) material removal due to grit throw and (b) material removal due to grit hammering (Shaw, 1956). The depth of penetration due to grit throw can be calculated by:

$$h_{th} = \pi\, a_t v\, d_a \sqrt{\frac{\rho_a^3}{6\sigma_w}} \qquad (14.1)$$

where h_{th} is the depth of penetration due to grit throw, mm; a_t is 2 times the amplitude of tool oscillation, mm; v is the frequency of the tool oscillation, Hz; ρ_a is the density of abrasive grit, g/mm³; and σ_w is the mean stress acting on the work surface, MPa (see Example 14.1).

Now, the MRR due to grit throw can be computed by:

$$MRR_{th} = k_1 \cdot k_2 \cdot k_3 \left(\sqrt{\frac{v\, h_{th}^3}{d_a}} \right) \tag{14.2}$$

where MRR_{th} is the MRR due to grit throw, mm³/s; and k_1, k_2, k_3 are the constants of proportionality (see Example 14.2).

The depth of penetration due to grit hammering can be calculated by:

$$h_h \cong h_w = \sqrt{\frac{4\, F_f\, a_t\, d_a}{\pi\, k_2 \left(j+1 \right) \sigma_w}} \tag{14.3}$$

where h_w is the depth of workpiece penetration due to grit hammering, mm; F_f is the feed force on the tool, N; and j is the ratio of the mean stress on the work to the mean stress acting on the tool ($j = \dfrac{\sigma_w}{\sigma_t}$) (see Example 14.3).

The MRR by grit hammering (MRR_h) can be computed by:

$$MRR_h = k_1\, k_2\, k_3\, v\, \sqrt{\frac{h_h^3}{d_a}} \tag{14.4}$$

where h_h is the depth of penetration due to grit hammering (see Example 14.4).

Thus, the total MRR can be determined by (El-Hofy, 2005):

$$MRR = MRR_{th} + MRR_h \tag{14.5}$$

where MRR is the total material removal rate, mm³/s. The machining time for producing a hole by ultrasonic machining can be computed by:

$$t_m = \frac{\text{Volume of hole}}{MRR} = \frac{\frac{\pi}{4} D^2 t}{MRR} = \frac{\pi D^2 t}{4 \left(MRR \right)} \tag{14.6}$$

where t_m is the machining time, s; D is the hole diameter, mm; and t is the work thickness, mm; and MRR is the material removal rate, mm³/s (see Example 14.5).

14.3 ABRASIVE JET MACHINING (AJM)

14.3.1 AJM PROCESS AND ITS ADVANTAGES

In the preceding section, we learnt that *USM* is capable of machining brittle materials but the equipment cost is high; this limitation has been overcome by the development of abrasive jet machining (AJM). AJM involves the use of abrasive particles, propelled by a high-velocity gas, that strike on the workpiece and remove the material by micro-cutting action as well as brittle fracture of the work material

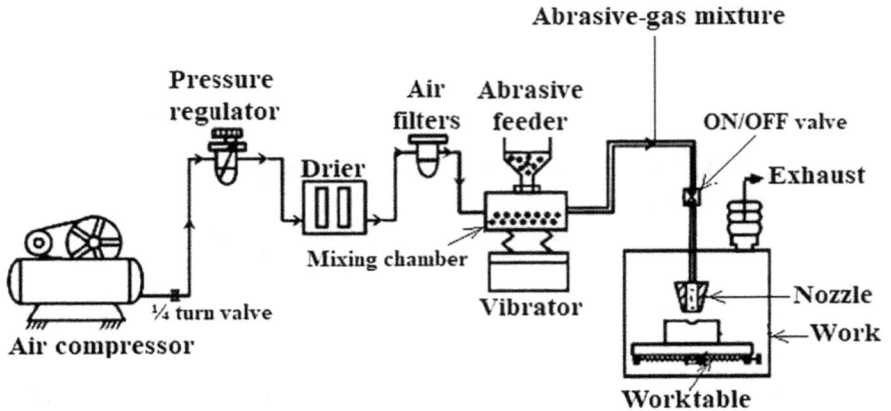

FIGURE 14.2 Abrasive jet machining

(see Figure 14.2). AJM process is commonly used to machine heat-sensitive, brittle, or hard materials. The distinct advantages of AJM include (a) fast equipment set up, (b) low equipment cost, (c) less vibration, (d) no thermally induced stresses in the workpiece, and (e) environmentally friendly process. Some limitations of AJM include low MRR, abrasive particles cannot be re-used, and the like.

14.3.2 AJM Process Parameters – Achieving Higher MRR/Better Surface Finish

The performance of AJM process strongly depends on the following process parameters: (a) nozzle tip distance, (b) abrasive mass flow rate, (c) gas pressure, (d) velocity of the abrasive particles, (e) abrasive-gas mixing ratio, and (f) abrasive grain size. The optimum ranges for some important process parameters are presented in Table 14.1. In AJM, the abrasive particles (generally with grain size of around 50 µm) impinge on the work material at a velocity of 200 m/s from the nozzle of internal diameter (ID) of 0.5 mm with a nozzle-tip distance of around 1.6 mm.

The abrasive particle size plays an important role in controlling the MRR as well as surface finish. In order to achieve a high MRR, a coarser grit (particle size ≈ 70 µm) is recommended. On the other hand, a finer grit (particle size ≈ 30 µm) is selected for a good surface finish in the work material. The nozzle-tip distance is also

TABLE 14.1
The Optimum Process Parameters for AJM

Process Parameter	Abrasive Grain Size	Abrasive Velocity	Nozzle Tip Distance	Nozzle ID
Values range	30–70 µm	150–300 m/s	0.75–2 mm	0.2–0.8 mm

an important process parameter in AJM. The MRR increases with the nozzle-tip distance up to a certain limit. After that limit, the MRR remains constant to some extent and then decreases (see Table 14.1). For brittle materials, higher MRR would occur with close to 90° impingement (see Figure 14.2), whereas for ductile materials, lower impingement angles (such as 30°) would ensure the highest MRR.

14.3.3 Mathematical Modeling of AJM

In AJM, the ratio of the mass flow rate of abrasive particles to the mass flow rate of the abrasive and the carrier gas is called the *mixing ratio*. Mathematically,

$$\text{Mixing ratio} = \frac{\textit{Mass flow rate of abrasive particle}}{\textit{mass flow rate of abrasive and carrier gas}} = \frac{\dot{m}_a}{\dot{m}_{a+g}} \quad (14.7)$$

The mixing ratio can be increased by increasing mass flow rate of abrasive grits maintaining carrier gas flow rate constant. The MRR can be improved by proportionally increasing both the abrasive flow rate and gas flow rate at same rate maintaining mixing ratio constant (see Example 14.6).

The MRR, in AJM, strongly depends on the following factors: (a) abrasive mass flow rate, (b) abrasive velocity, (c) density of abrasive grit, and (d) flow strength of the work material. Accordingly, the MRR for a brittle material can be calculated by:

$$MRR_{brittle} = \frac{\dot{m}_a v^{1.5}}{\rho_a^{0.25} \sigma_w^{0.75}} \quad (14.8)$$

where $MRR_{brittle}$ is MRR of the brittle material being machined, mm³/s; $\dot{m}_a$ is mass flow rate of abrasives, kg/s; v is velocity of abrasive grits at the striking point, mm/s; ρ_a is density of each abrasive particle, kg/mm³; and σ_w is flow stress on the work material, MPa.

The MRR for a ductile material can be calculated by (Jain, 2007):

$$MRR_{ductile} = 0.5 \left(\frac{\dot{m}_a v^2}{\sigma_w} \right) \quad (14.9)$$

where $MRR_{ductile}$ is the MRR for the ductile material being machined, mm³/s.

The significance of Equations 14.8–14.9 is illustrated in Examples 14.7–14.8.

14.4 WATER JET MACHINING

14.4.1 Water Jet Machining Process and Its Advantages

Water Jet Machining (WJM) is a mechanical-energy-based NTM process that involves the use of ultra-high-pressure high-velocity water jet to cut and machine soft and polymeric materials (such as foam, plastic, etc.). In *WJM*, water from the fluid supply is pumped to the intensifier using a hydraulic pump, which increases the pressure of the water to the required level (~350 MPa). Then the pressurized

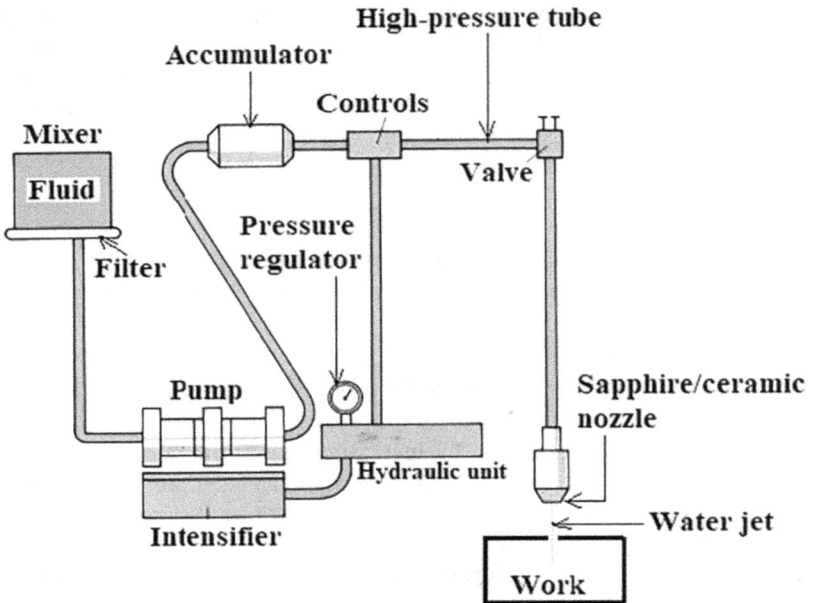

FIGURE 14.3 Water jet machining (WJM)

water is passed to an accumulator where pressurized water is temporarily stored (see Figure 14.3). The pressurized water then passes through a nozzle by use of control valve and flow regulator. At the nozzle, there is a tremendous increase in the kinetic energy of the pressurized water. As the high-energy water jet strikes the workpiece, stresses are induced leading to deformation and the removal of material from the workpiece. The WJM process is capable of cutting softer materials in a sandwich up to 10-cm thickness. It is also employed in cutting aluminum foils in food processing industries.

14.4.2 Mathematical Modeling of Water Jet Machining

In WJM, the water jet velocity (immediately after issuing from the nozzle) can be calculated by:

$$v_w = C\sqrt{\frac{2P}{\rho_w}} = 0.85\sqrt{\frac{2P}{\rho_w}} \qquad (14.10)$$

where v_w is the water jet velocity, m/s; P is the water pressure, Pa; and ρ_w is the density of water ($\rho_w = 1000$ kg/m^3); and C is the flow rate coefficient ($C = 0.85$).

The mass flow rate of water being issued (as a jet) from the nozzle orifice can be computed by:

$$\dot{m}_w = \frac{\pi}{4}\rho_w d_o^2 v_w \qquad (14.11)$$

TABLE 14.2
Atmospheric Deceleration Coefficients for Various Nozzle-tip Distances in WJM

Nozzle-tip distance (mm)	200	300	500
Atmospheric deceleration coefficient, C_v	1	0.95	0.85

where $\dot{m}_w$ is the mass flow rate, kg/s, and d_o is the orifice diameter, m (see Example 14.9). The impact force exerted by the water jet on the workpiece can be calculated by:

$$F_{wji} = \dot{m}_w \, C_v \, v_w \qquad (14.12)$$

where F_{wji} is the water jet impact force, N; $\dot{m}_w$ is the mass flow rate of water, kg/s; C_v is the atmospheric deceleration coefficient dependent on the nozzle-tip distance (see Table 14.2), and v_w is the water jet velocity (immediately after the nozzle), m/s (see Example 14.10).

14.5 ABRASIVE WATER JET MACHINING (AWJM)

14.5.1 AWJM PROCESS AND ITS ADVANTAGES

In the preceding section, we learnt that clean and safe cuts can be made at a low cost by using WJM process; however, the process is unsuitable of machining steels and other metals. This difficulty is overcome by abrasive water jet machining (AWJM). AWJM involves the use of a fine jet of very high-pressure water and abrasive slurry to more effectively cut the target material (including steels and other metals) by means of erosion (Gupta et al., 2014). The working principle of AWJM is similar to WJM except that an abrasive slurry is mixed with water in AWJM (see Figure 14.4). The commonly used abrasive is red garnet. In order to achieve a higher MRR, a coarser abrasive with a lower grit number in the range of 60–80 should be used. On the other hand, a finer abrasive with a higher grit number in the range of 100–150 is recommended to obtain a better surface finish (lower surface roughness, R_a). The surface roughness of the material machined by AWJM depends on several factors,

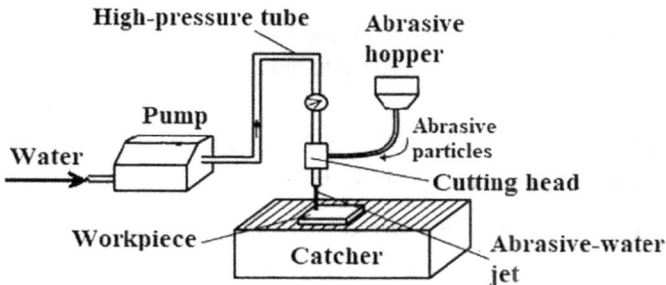

FIGURE 14.4   Schematic of Abrasive water jet machining

including the abrasive grit size, nozzle-tip distance, water jet pressure, workpiece thickness, abrasive flow rate, and the feed rate. A mathematical expression to estimate the surface roughness of the material machined by AWJM is presented in the following sub-section.

14.5.2 MATHEMATICAL MODELING OF AWJM

In the preceding section, a mathematical model for determining the mass flow rate of water in WJM has been presented; this model (Equation 14.11) is also applicable to AWJM. The ratio of the mass flow rate of abrasives to the mass flow rate of water, R, is given by:

$$R = \frac{Mass\ flow\ rate\ of\ abrasives}{Mass\ flow\ rate\ of\ water} = \frac{\dot{m}_a}{\dot{m}_w} \tag{14.13}$$

Now, the abrasive water jet velocity can be calculated by:

$$v_{awj} = \left(\frac{1}{1+R}\right)v_{wj} \tag{14.14}$$

where v_{wj} is the water jet velocity, m/s (see Example 14.11).

The depth of penetration of the abrasive water jet into the workpiece can be calculated by:

$$h_{AWJM} = \frac{\pi}{4}d_o^2\,\text{R}\left(\frac{1}{1+R}\right)^2 \frac{P^{1.5}}{U_{work}\,d_i\,V_t}\sqrt{\frac{2}{\rho_w}} \tag{14.15}$$

where h_{AWJM} is the depth of penetration in AWJM, m; d_o is the nozzle orifice diameter, m; P is the water pressure, Pa; U_{work} is the specific energy of the work material, J/m³; d_i is the insert diameter, m; and V_t is the traverse speed of the workpiece, m/s (see Example 14.12).

The surface roughness of AWJM-machined material can be computed by (Janković et al., 2012):

$$R_a = 0.2913\left(\frac{t^{0.694}\,f_r^{0.648}}{q^{0.212}}\right) \tag{14.16}$$

where R_a is the surface roughness, μm; t is the workpiece thickness, mm; f_r is the feed rate, mm/min; and q is the abrasive flow rate, g/min (see Example 14.13).

14.6 CALCULATIONS – *WORKED EXAMPLES ON MECHANICAL-ENERGY MACHINING*

EXAMPLE 14.1: CALCULATING THE DEPTH OF PENETRATION DUE TO GRIT THROW IN USM

A 5-mm-diameter hole is produced in a tungsten carbide plate of thickness 1½ times the hole diameter by the USM process. The mean stress acting on the work surface is

6900 MPa. The mean abrasive grain diameter is 14 μm. The feed force on the tool is 3 N. The frequency of tool oscillation is 24 kHz and the amplitude of tool vibration is 24 μm. The mean stress acting on the tool is 1300 MPa. The abrasive density is 4 g/cm^3. The values of constants are $k_1 = 0.3$, $k_2 = 1.8$ mm^2, and $k_3 = 0.6$. Calculate the depth of penetration due to grit throw in the USM process.

Solution

$a_t = 2 \times$ amplitude $= 2 \times 24$ μm $= 48 \times 10^{-6}$ m $= 0.048$ mm; $v = 24$ kHz $= 24{,}000$ Hz
$d_a = 14$ μm $= 14 \times 10^{-6}$ m $= 0.014$ mm; $\rho_a = 4$ g/cm^3 $= 0.004$ g/mm^3; $\sigma_w = 6900$ MPa
By using Equation 14.1,

$$h_{th} = \pi\, a_t\, v\, d_a \sqrt{\frac{\rho_a^3}{6\sigma_w}} = \pi \times 0.048 \times 24000 \times 0.014 \sqrt{\frac{0.004^3}{6 \times 6900}} = 6.3 \times 10^{-5}\,\text{mm}$$

The depth of penetration due to grit throw $= h_{th} = 6.3 \times 10^{-5}$ mm

EXAMPLE 14.2: COMPUTING THE MRR DUE TO GRIT THROW IN USM

By using the data in Example 14.1, compute the MRR due to grit throw.

Solution

$v = 24{,}000$ Hz, $h_{th} = 6.3 \times 10^{-5}$ mm, $h_{th}^3 = 25 \times 10^{-14}$ mm^3, $d_a = 0.014$ mm
By using Equation 14.2,

$$MRR_{th} = k_1 \cdot k_2 \cdot k_3 \left(\sqrt{\frac{v\, h_{th}^3}{d_a}}\right) = (0.3 \times 1.8 \times 0.6)\left(\sqrt{\frac{24000 \times 25 \times 10^{-14}}{0.014}}\right)$$

$$MRR_{th} = 0.324 \times \sqrt{42857142.8 \times 10^{-14}} = 0.324 \times 6546.54 \times 10^{-7} = 2.12 \times 10^{-4}\,\text{mm}^3/\text{s}$$

The MRR due to grit throw $= MRR_{th} = 2.12 \times 10^{-4}$ mm^3/s

EXAMPLE 14.3: CALCULATING THE DEPTH OF PENETRATION DUE TO GRIT HAMMERING IN USM

By using the data in Example 14.1, calculate the depth of penetration due to grit hammering in the USM process.

Solution

$F_f = 3$ N; $a_t = 0.048$ mm; $d_a = 0.014$ mm; $\sigma_w = 6900$ MPa; $\sigma_t = 1{,}300$ MPa, $k_2 = 1.8$ mm^2

$$j = \frac{\sigma_w}{\sigma_t} = \frac{6900}{1300} = 5.3$$

By using Equation 14.3,

$$h_h \cong h_w = \sqrt{\frac{4 F_f\, a_t\, d_a}{\pi k_2 (j+1)\sigma_w}} = \sqrt{\frac{4 \times 3 \times 0.048 \times 0.014}{\pi \times 1.8 \times 6900(5.3+1)}} = \sqrt{\frac{0.0081}{245817}} = 1.8 \times 10^{-4}\,\text{mm}$$

The depth of penetration due to grit hammering = $h_h \cong 1.8 \times 10^{-4}$ mm

EXAMPLE 14.4: CALCULATING THE MRR BY GRIT HAMMERING IN USM

By using the data in Example 14.3, calculate the MRR by grit hammering in the USM process.

Solution
$k_1 = 0.3$, $k_2 = 1.8$ mm^2, $k_3 = 0.6$; $v = 24{,}000$ Hz; $d_a = 0.014$ mm; $h_h = 1.8 \times 10^{-4}$ mm
By using Equation 14.4,

$$MRR_h = k_1 k_2 k_3 v \sqrt{\frac{h_h^3}{d_a}} = \left(0.3 \times 1.8 \times 0.6 \times 24{,}000\right)\sqrt{\frac{\left(1.8 \times 10^{-4}\right)^3}{0.014}}$$

$$= 7776 \times \sqrt{\frac{\left(0.00018\right)^3}{0.014}}$$

$$MRR_h = 7776 \times 2.04 \times 10^{-5} = 0.1587 \, \text{mm}^3/\text{s}$$

EXAMPLE 14.5: COMPUTING THE MACHINING TIME FOR A USM OPERATION

By using the data in Examples 4.1–4.4, calculate the machining time for the USM operation.

Solution
$MRR_{th} = 0.000212$ mm^3/s; $MRR_h = 0.158$ mm^3/s; $D = 5$ mm; $t = 1.5$ D = $1.5 \times 5 = 7.5$ mm
By using Equation 14.5,

$$MRR = MRR_{th} + MRR_h = 0.000212 + 0.158 = 0.158212 \, \text{mm}$$

By using Equation 14.6,

$$t_m = \frac{\pi D^2 t}{4(MRR)} = \frac{3.142 \times 5^2 \times 7.5}{4 \times 0.158212} = 931 \, \text{s} = 15.5 \, \text{min}$$

The machining time for the USM operation = 15.5 min

EXAMPLE 14.6: CALCULATING THE MASS OF THE ABRASIVE PARTICLES REQUIRED IN AJM

The mass flow rate of the abrasive and the carrier gas in an AJM process is 3.3 g/min. It is desired to keep the mixing ratio constant as 1.5. What mass of the abrasive particles is required if the AJM process takes 10 min?

Solution

$\dot{m}_{a+g}$ = 3.3 g/min, Mixing ratio = 1.5,

By using the modified form of Equation 14.7,

$$\dot{m}_a = \left(\dot{m}_{a+g}\right)\left(\text{mixing ratio}\right) = 3.3 \times 1.5 = 4.95\,\text{g/min}$$

$$m_a = \left(\dot{m}_a\right)\left(\text{time}\right) = 4.95 \times 10 = 49.5\,\text{g}.$$

The mass of the abrasive particles required = 49.5 g.

EXAMPLE 14.7: CALCULATING THE MRR FOR MACHINING A BRITTLE MATERIAL BY AJM PROCESS

The flow stress on a brittle material during AJM process is 3.2 GPa. The abrasive mass flow rate is 4 g/min, and the velocity of the abrasive grit at the striking point is 260 m/s. The density of the abrasive grit is 2.9 g/cm³. Compute the MRR in the AJM process.

Solution

$\dot{m}_a$ = 4 g/min = 4 × 0.001 kg/60 s = 66.667 × 10⁻⁶ kg/s
v = 260 m/s = 260 × 10³ mm/s = 260,000 mm/s; σ_w = 3.2 GPa = 3.2 × 10³ MPa = 3200 MPa
ρ_a = 2.9 g/cm³ = 2.9 × 0.001 kg/(10 mm)³ = 2.9 × 10⁻³ kg/10³ mm³ = 2.9 × 10⁻⁶ kg/mm³

By using Equation 14.8,

$$MRR_{brittle} = \frac{\dot{m}_a v^{1.5}}{\rho_a^{0.25} \sigma_w^{0.75}} = \frac{66.667 \times 10^{-6} \times (260000)^{1.5}}{(2.9 \times 10^{-6})^{0.25} \times 3200^{0.75}} = \frac{8838.34}{0.041266 \times 425.463}$$
$$= 503.4\,\text{mm}^3/\text{s}$$

The MRR = 503.4 mm³/s.

EXAMPLE 14.8: CALCULATING THE MRR FOR A DUCTILE MATERIAL BY AJM

The flow strength of a ductile material to be machined by AJM is 200 MPa. The abrasive mass flow rate is 4 g/min, and the velocity of the abrasive grit at the striking point is 260 m/s. Compute the MRR for machining the ductile material by AJM.

Solution

$\dot{m}_a$ = 4 g/min = 4 × 0.001 kg/60 s = 66.667 × 10⁻⁶ kg/s
v = 260 m/s = 260 × 10³ mm/s = 260,000 mm/s; σ_w = 200 MPa

By using Equation 14.9,

$$MRR_{ductile} = 0.5\left(\frac{\dot{m}_a v^2}{\sigma_w}\right) = \frac{0.5 \times 66.667 \times 10^{-6} \times 260000^2}{200} = 0.1667 \times 6.76 \times 10^4$$

$$MRR_{ductile} = 1.11 \times 10^4 \text{ mm}^3/\text{s}$$

EXAMPLE 14.9: CALCULATING THE MASS FLOW RATE OF WATER IN WJM

The nozzle orifice diameter in a WJM equipment is 0.3 mm. The equipment is used to perform WJM at a water pressure of 380 MPa. Calculate the mass flow rate of water in the water jet.

Solution
$d_o = 0.3$ mm $= 0.0003$ m, $P = 380$ MPa $= 380 \times 10^6 = 3.8 \times 10^8$ Pa, $\rho_w = 1000$ kg/m³
  By using Equation 14.10,

$$v_{wj} = 0.85 \times \sqrt{\frac{2P}{\rho_w}} = 0.85 \times \sqrt{\frac{2 \times 3.8 \times 10^8}{1000}} = 0.85 \times 871.78 = 741 \text{ m/s}$$

By using Equation 14.11,

$$\dot{m}_w = \frac{\pi}{4}\rho_w d_o^2 v_w = \frac{\pi}{4} \times 1000 \times 0.0003^2 \times 741 = 0.0524 \text{ kg/s}$$

The mass flow rate of water in the water jet $= \dot{m}_w = 0.0524$ kg/s $= 3.143$ kg/min.

EXAMPLE 14.10: CALCULATING THE WATER JET IMPACT FORCE IN WJM

By using the data in Example 14.9, calculate the impact force exerted by the water jet on the workpiece in the AJM process when the nozzle-tip distance is 300 mm.

Solution
$C_v = 0.95$ (see Table 16.2), $\dot{m}_w = 0.0524$ kg/s, $v_w = 741$ m/s
  By using Equation 4.12,

$$F_{wji} = \dot{m}_w C_v v_w = 0.0524 \times 0.95 \times 741 = 36.887 \text{ N}$$

The water jet impact force $= 36.89$ N

EXAMPLE 14.11: CALCULATING THE ABRASIVE WATER JET VELOCITY IN AWJM

The nozzle orifice diameter in an AWJM equipment is 0.3 mm. The equipment is used to perform AWJM to machine a steel workpiece at a traverse speed of 320 mm/min with an insert diameter of 1 mm. The mass flow rate of abrasive is 0.018 kg/s, and the water pressure is 380 MPa. Calculate the abrasive water jet velocity in the AWJM process.

Solution

$P = 380$ MPa, $d_o = 0.3$ mm, $\dot{m}_a = 0.018$ kg/s, $v_{AWJ} = $?
By reference to Example 14.9, $v_{wj} = 741$ m/s, $\dot{m}_w = 0.0524$ kg/s.
By using Equation 14.13,

$$R = \frac{\dot{m}_a}{\dot{m}_w} = \frac{0.0180}{0.0524} = 0.3435$$

By using Equation 14.14,

$$v_{awj} = \left(\frac{1}{1+R}\right)v_{wj} = \left(\frac{1}{1+0.3435}\right) \times 741 = 551.5 \, \text{m/s}$$

The abrasive water jet velocity = 551.5 m/s

EXAMPLE 14.12: CALCULATING THE DEPTH OF PENETRATION IN AN AWJM PROCESS

By using the data in Example 14.11, calculate the depth of penetration of the abrasive water jet into the workpiece. The specific energy of steel is 13.6 J/mm³

Solution

$R = 0.3435$, $d_o = 0.3$ mm = 0.0003 m, $P = 380$ MPa = $380 \times 10^6 = 3.8 \times 10^8$ Pa
$d_i = 0.001$ m, $V_t = 320$ mm/min = 5.3×10^{-3} m/s, $\rho_w = 1000$ kg/m³; $U_{work} = 13.6 \times 10^9$ J/m³
By using Equation 14.15,

$$h_{AWJM} = \frac{\pi}{4} d_o^2 R \left(\frac{1}{1+R}\right)^2 \frac{P^{1.5}}{U_{work} d_i V_t} \sqrt{\frac{2}{\rho_w}}$$

$$h_{AWJM} = \frac{\pi}{4} \times (0.0003)^2 \times 0.343 \times \left(\frac{1}{1+0.343}\right)^2 \frac{\left(3.8 \times 10^8\right)^{1.5}}{13.6 \times 10^9 \times 0.001 \times 5.3 \times 10^{-3}} \sqrt{\frac{2}{1000}}$$

$$h_{AWJM} = 2.424 \times 10^{-8} \times 0.554 \times \frac{3.8^{1.5} \times 10^{12}}{0.0721 \times 10^6} \times 0.0447 = 0.06 \times 10^{-8} \times \frac{7.41 \times 10^6}{0.0721}$$

$$h_{AWJM} = 6.166 \times 10^{-2} \, \text{m} = 6.166 \times 10^{-2} \times 1000 = 61.6 \, \text{mm}$$

The depth of penetration in the AWJM process = 61.6 mm

EXAMPLE 14.13: ESTIMATING THE SURFACE ROUGHNESS OF AN AWJM MACHINED MATERIAL

The abrasive flow rate in an AWJM process is 300 g/min. The workpiece plate thickness is 6.8 mm, and the feed rate is 700 mm/min. Estimate the surface roughness of the plate after machining.

Solution

$t = 6.8$ mm, $f_r = 700$ mm/min, $q = 300$ g/min

By using Equation 14.16,

$$R_a = 0.2913\left(\frac{t^{0.694} f_r^{0.648}}{q^{0.212}}\right) = 0.2913\left(\frac{(6.8)^{0.694}(700)^{0.648}}{300^{0.212}}\right) = 0.2913\left(\frac{3.78\times69.76}{3.35}\right)$$

$$= 23\,\mu m$$

QUESTIONS AND PROBLEMS

14.1   a. Briefly explain the ultrasonic machining (USM) process with the aid of sketches.

       b. What are the advantages and limitations of USM?

14.2   a. Diagrammatically illustrate the abrasive jet machining (AJM) process.

       b. List the advantages and limitations of AJM.

       c. Give the typical process parameter values generally practiced in AJM.

       d. What are techniques to achieve higher material removal rates (MRR) in AJM?

14.3   Briefly explain water jet machining (WJM) with the aid of a diagram.

14.4   Diagrammatically illustrate the abrasive water jet machining (AWJM) process.

P14.5   The mass flow rate of the abrasive and the carrier gas, in an AJM process, is 2.8 g/min. It is desired to keep the mass ratio constant as 1.3. What mass of the abrasive particles is required if the AJM process takes 12 min?

P14.6   A 4-mm-diameter hole is produced in a tungsten carbide plate having thickness double. The hole diameter by USM process. The mean stress acting on the work surface is 7000 MPa. The mean abrasive grain diameter is 15 μm. The feed force on the tool is 3.2 N. The frequency of tool oscillation is 25 kHz and the amplitude of tool vibration is 23 μm. The mean stress acting on the tool is 1200 MPa. The abrasive density is 4.2 g/cm^3. The values of proportionality constants are: $k_1 = 0.3$, $k_2 = 1.8$ mm^2, $k_3 = 0.6$. Calculate the (a) material removal rate in the USM process and (b) machining time for the process.

P14.7   The flow stress on a brittle material, during AJM process, is 3 GPa. The abrasive mass flow rate is 3.8 g/min, and the velocity of the abrasive grit at the striking point is 300 m/s. The density of the abrasive grit is 3.5 g/cm^3. Compute the material removal rate.

P14.8   The nozzle orifice diameter, in a WJM equipment, is 0.32 mm. The equipment is used to perform WJM at a water pressure of 400 MPa. Calculate the impact force exerted by the water jet on the workpiece in the AJM process when the nozzle tip distance is 500 mm.

P14.9   The abrasive flow rate in an AWJM process is 320 g/min. The workpiece plate thickness is 5.5 mm, and the feed rate is 650 mm/min. Estimate the surface roughness of the plate after machining.

P14.10 The nozzle orifice diameter, in AWJM equipment, is 0.3 mm. The equipment is used to perform AWJM to machine a steel workpiece at a traverse speed of 290 mm/min with an insert diameter of 1 mm. The mass flow rate of abrasive is 0.02 kg/s, and the water pressure is 400 MPa. Calculate the depth of penetration of the abrasive water jet into the workpiece. The specific energy of steel is 13.6 J/mm³.

MCQS

14.11 Encircle the most appropriate answers for the following statements?
   a. Which NTM process is incapable of machining alloys or ceramics?
      (i) USM, (ii) WJM, (iii) AWJM, (iv) AJM
   b. Which NTM equipment involves the use of a transducer?
      (i) USM, (ii) WJM, (iii) AWJM, (iv) AJM
   c. Which NTM process does not require use of abrasive grits?
      (i) USM, (ii) WJM, (iii) AWJM, (iv) AJM
   d. Which NTM process is the best in terms of equipment cost and its set up?
      (i) USM, (ii) WJM, (iii) AWJM, (iv) AJM

REFERENCES

El-Hofy, H. A-G. (2005). *Advanced Machining Processes: Nontraditional and Hybrid Machining Processes*, McGraw-Hill Education Inc., New York.

Grzesik, W. (2016). *Advanced Machining Processes of Metallic Materials* (2nd Edition), Elsevier Science Publications Inc, New York.

Gupta, V., Pandey, P.M., Garg, M.P., Khanna, R. Batra, N.K. 2014. Minimization of *kerf taper* angle and kerf width using Taguchi's Method in abrasive water jet machining of marble, *Procedia Material Science*, 6, 140–149.

He, J., Guo, Z., Qin, Y., Chen, X. (2018), Reducing edge chipping of micro-holes with electro-phoresis-assisted micro-*ultrasonic machining*, *Procedia CIRP*, 682018, 444–449.

Jain, V.K. (2007), *Advanced Machining Processes*, Allied Publishers Pvt. Ltd., New Delhi.

Janković, P. Igić, T., Nikodijević, D. 2012. Process parameter effect on material removal mechanism and cut quality of abrasive water jet machining. *Theoretical and Applied Mechanics Series: Special Issue - Address to Mechanics*, 40(S1), 277–291, Belgrade.

Kuo, K.L., Hocheng, H., Hsu, C.C. (2013), Ultrasonic machining, In: Hocheng, Hong, Tsai, Hung-Yin (Eds.), *Advanced Analysis of Non-traditional Machining*, Springer, Germany.

Shaw, M.C. (1956), Ultrasonic grinding, *Microtechnic*, 10(6), 257–265.

15 Thermal Energy-based Machining Processes

15.1 WHAT ARE THERMAL ENERGY MACHINING PROCESSES?

We learnt in the preceding chapter that non-traditional machining (NTM) processes are generally classified into four groups: (1) mechanical energy-based machining processes, (2) thermal energy-based machining processes, (3) electrochemical machining processes, and (4) chemical machining. Thermal energy NTM, or thermal energy machining, involves the application of thermal energy to a very small portion of the work surface, causing the portion to be removed by fusion and/or vaporization. Most of these processes involve the conversion of electrical energy into thermal energy for material removal. There are four commonly used types of thermal energy machining processes: (a) electric discharge machining (*EDM*), (b) electric discharge wire cutting (*EDWC*), (c) electron beam machining (*EBM*), (d) laser beam machining (*LBM*), and (e) plasma arc cutting (PAC); these processes are discussed in the following sections.

15.2 ELECTRIC DISCHARGE MACHINING (EDM)

15.2.1 THE EDM PROCESS AND ITS APPLICATIONS

Electric discharge machining (*EDM*) involves the removal of metal by means of electric spark that creates localized temperatures high enough to melt or vaporize the metal. In *EDM*, a dielectric liquid is subjected to an electric voltage by using two electrodes resulting in material removal from the metallic workpiece due to a series of rapidly recurring current discharges between the two electrodes (Jahan, 2015). The tool electrode is made cathode (−), while the work electrode is anode (+), as shown in Figure 15.1. The discharges (sparks) are generated by a pulsating direct current power supply connected to the work and the tool. The discharge occurs at the location where the two surfaces are the closest. The dielectric fluid (e.g. kerosene oil) ionizes at this location to create a path for the discharge. The small region in which discharge occurs is heated to very high temperatures causing melting and material removal. The flowing dielectric carries away the small particle or "debris". After removing the debris, the tool is fed to keep the gap between the tool and work constant; the short gap makes possible striking of next discharge(s) between the electrodes.

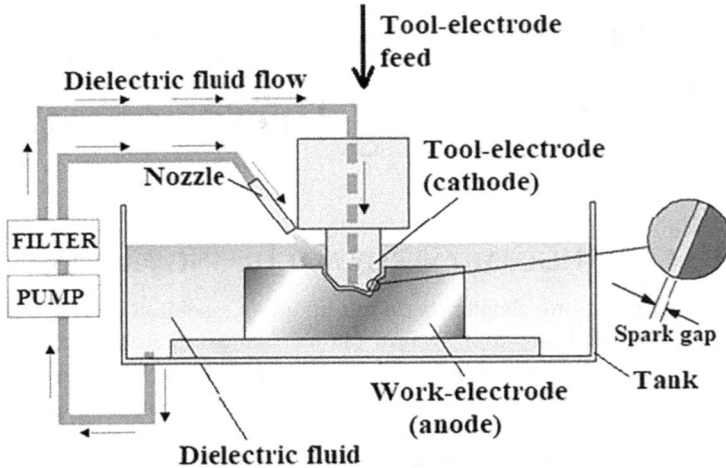

FIGURE 15.1   Electric discharge machining (EDM)

The EDM process is suitable for manufacturing cutting tools and metallic molds and dies, such as molds for plastic injection molding, extrusion dies, wire-drawing dies, forging and heading dies, and the like.

15.2.2 MATHEMATICAL MODELING OF EDM

In *EDM*, the set of tool electrode and the work electrode may be considered as a capacitor due to the use of dielectric fluid in between them. The electrical energy required to create a discharge (spark) is called the *spark energy*. The *spark energy* can be computed by:

$$E_s = \frac{C V^2}{2} \tag{15.1}$$

where E_s is the spark energy, J; C is the capacitance of the capacitor, F; and V is the maximum voltage applied, volts (see Example 15.1).

The *MRR* in EDM can be estimated by:

$$\text{MRR} = 40,000 \, I \, T_w^{-1.23} \tag{15.2}$$

where *MRR* is the metal removal rate, mm³/min; I is the current, A; and T_w is the melting temperature of the work material, °C (see Example 15.2).

15.3 ELECTRIC DISCHARGE WIRE CUTTING (*EDWC*)

15.3.1 THE EDWC PROCESS AND ITS ADVANTAGES

Electric discharge wire cutting (*EDWC*), also called *wire EDM*, involves the use of a continuously spooling conductive wire (cathode) and a dielectric fluid (e.g.

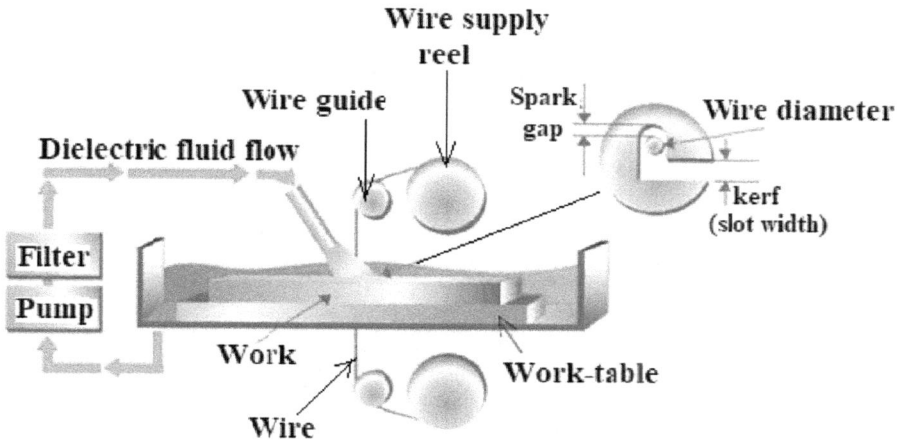

FIGURE 15.2 Electric discharge wire cutting (EDWC)

deionized water) for EDM of a workpiece (anode). A power supply generates rapid electric pulses that result in discharges between the work (anode) and the wire (cathode). The dielectric fluid submerging the work-part is directed to the tool–work interface using a nozzle. Thus the tool (wire) discharges thousands of sparks to the metal workpiece, thereby causing the melting or vaporization of minute pieces of metal resulting in material removal (see Figure 15.2). The workpiece is fed continuously and slowly past the wire in order to achieve the desired cutting path. The wire is usually made of copper or brass and is typically about 250 μm in diameter thereby enabling the operator to make narrow cuts/slots.

The EDWC process is used to cut plates as thick as 30 cm and for making punches, tools, and dies made of hard alloys. It is considered as an efficient and economical tool in machining of electrically conductive advanced alloys and composites (Owhal, 2020).

15.3.2 Mathematical Modeling of EDWC

In EDWC, the kerf or slot width (b, mm) can be calculated by:

$$b = d_w + 2s \tag{15.3}$$

where d_w is the wire diameter, mm, and s is the spark gap, mm (see Example 15.4)
The *MRR* in EDWC can be calculated by:

$$MRR = f_r h\, b \tag{15.4}$$

where *MRR* is the metal removal rate, mm³/min; f_r is the feed rate, mm/min; h is the height or thickness, mm; and b is the slot width, mm (see Example 15.5).

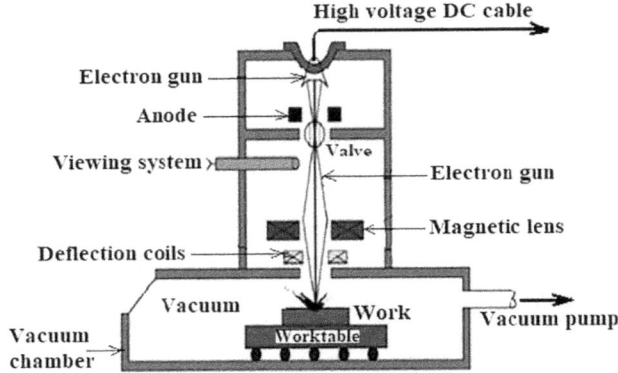

FIGURE 15.3   Electron beam machining (EBM)

15.4 ELECTRON BEAM MACHINING (EBM)

15.4.1 THE EBM PROCESS

Electron beam machining (EBM) employs high-energy electrons to cut small slots or drill very small diameter holes. In *EBM*, firstly a workpiece is placed in a vacuum chamber (pressure = 10^{-4} torr). When the electron gun (with a tungsten filament) is connected to a high voltage (~150 kV), the tungsten filament is heated up to a temperature above 2500°C. The presence of anode causes ejection of electrons, which are focused by the field formed by the grid cup and by a magnetic lens system (see Figure 15.3). The high-velocity focused electrons strike workpiece over a small circular area thereby causing correspondingly rapid increase in the temperature of the workpiece, to well above its boiling point. This thermal effect results in material removal (Schneider, 1989; Okada, 2019). The *EBM* is capable to machine both soft and hard metallic and non-metallic materials. It is possible to drill small (0.1–0.3 mm diameter) holes with high depth-to-diameter ratios (> 100:1) by EBM.

15.4.2 MATHEMATICAL MODELING OF EBM

In *EBM*, the electrical energy of electron gun is first transferred to electron beam, and then the high kinetic energy of electron beam is translated into the heat energy of the workpiece. The electrical energy transferred to the electron beam can be given by:

$$E = e \cdot V \tag{15.5}$$

where E is the energy transferred, J; e is the charge on an electron, Coul.; and V is the potential difference between the filament and the anode, volts. The electron starts from rest (from the electron gun) so the kinetic energy gained by electrons is:

$$E_k = \tfrac{1}{2} \mathrm{m} \cdot v^2 \tag{15.6}$$

where E_k is the kinetic energy of electron beam, J; m is the mass of an electron, kg; and v is the speed of electron, m/s. Each electron gains kinetic energy equal to the amount of energy transferred from the electron gun. By combining Equations 15.5–15.6, we obtain:

$$\tfrac{1}{2}\,\mathrm{m}\cdot v^2 = e\cdot V \tag{15.7}$$

The significance of Equation 15.7 is illustrated in Example 15.6.

The depth of penetration of the electron beam in the work, in EBM, can be calculated by:

$$d = 2.6\times10^{-17}\left(\frac{V^2}{\rho}\right) \tag{15.8}$$

where d is the depth of penetration, mm; V is the accelerating voltage, volts; and ρ is the density of the work material, kg/mm^3 (see Example 15.7).

The material removal rate (MRR) in EBM can be determined by using Equation 3.9, while the unit power data is given in Table 3.1. Once the MRR has been calculated, the cutting speed, in EBM, can be calculated by:

$$v = \frac{\mathrm{MRR}}{A} \tag{15.9}$$

where v is the cutting speed, mm/s; MMR is the material removal rate, mm^3/s; and A is the area of the slot or hole to be machined, mm^2 (see Example 15.8).

15.5 LASER BEAM MACHINING (LBM)

Laser beam machining (*LBM*) involves the use of a high-energy laser (coherent light) beam to melt and vaporize particles on the surface of a metallic/non-metallic workpiece. In LBM, the laser beam is directed at the focus point on the workpiece surface, where the thermal energy is absorbed thereby transforming the work volume into a molten or vaporized state (Dahotre and Harimkar, 2008). The vaporized material is removed by flow of high-pressure assisted gas jet (see Figure 15.4).

15.6 PLASMA ARC CUTTING (PAC)

15.6.1 THE PAC PROCESS AND ITS ADVANTAGES/LIMITATIONS

The term "plasma" may be referred as "a superheated electrically ionized gas stream". Plasma arc cutting (PAC), also called *plasma fusion cutting*, is a thermal-energy non-traditional machining process that employs plasma funneled through a plasma torch to heat, melt and, ultimately cut electrically conductive thick materials. In *PAC*, an arc is formed between the electrode and the workpiece that is constricted by a fine-bore copper nozzle (Kumar et al., 2019). This increases

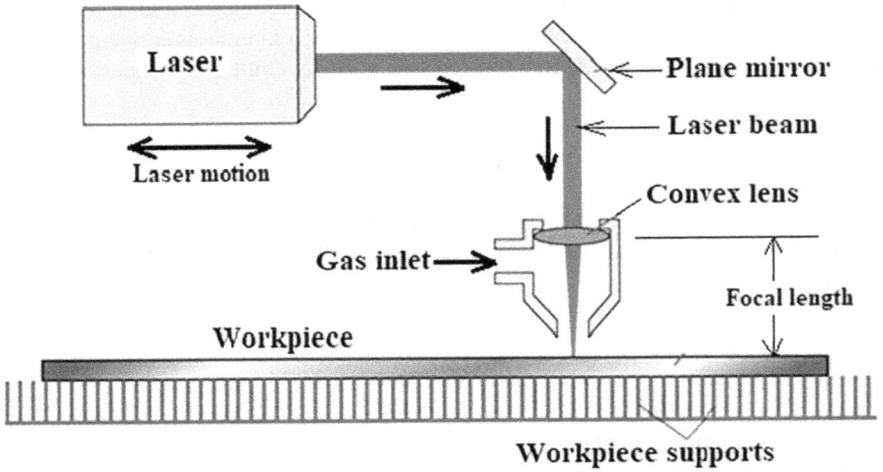

FIGURE 15.4   Laser beam machining (LBM)

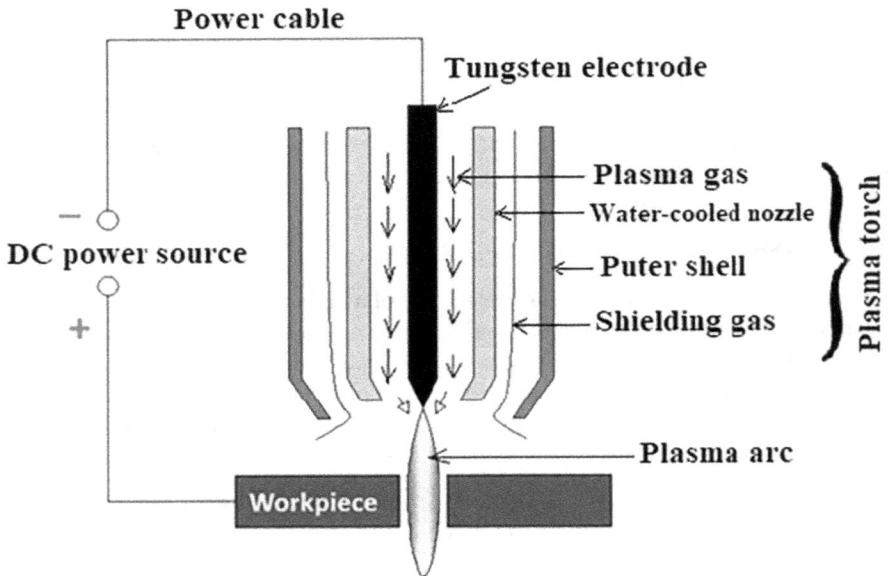

FIGURE 15.5   Plasma arc cutting (PAC)

the temperature and velocity of the plasma emanating from the nozzle. The temperature of the plasma is in excess of 15,000 °C, and the velocity can approach the speed of sound. The plasma gas flow rate is very high so the deeply penetrating plasma jet cuts through the thick material, and molten material is removed thereof (see Figure 15.5).

PAC is capable of cutting very hard and thick plates (up to 20 cm thickness). Most applications of PAC involve cutting of flat metal plates. The main limitations of PAC are (a) rough surface on cut edge and (b) metallurgical damage to cut surface due to excessively high temperature.

15.6.2 Mathematical Modeling for the Operating Cost in PAC

In PAC, the nozzle and the electrode are consumed during the process. The operating cost in PAC depends on a number of factors, including the PAC system (equipment) cost, the output current, the output voltage, the nozzle cost, the electrode cost, the duty cycle, the cutting speed, nozzle/electrode life (in arc h), and the like.

The consumable cost per arc hour in PAC can be calculated by:

$$Cost_{cons.\ per\ arc\ h} = \frac{C_n + C_e}{T_{L(cons.)}} \qquad (15.10)$$

where $Cost_{cons.\ per\ arc\ h}$ is the consumable cost per arc hour, \$/arc h; C_n is the nozzle cost, \$; C_e is the electrode cost, \$; and $T_{L(cons.)}$ is the nozzle life or electrode life, arc h.

The consumable cost per work hour can be calculated by:

$$Cost_{cons.\ per\ work\ h} = Cost_{cons.\ per\ arc\ h} * duty\,(\%) \qquad (15.11)$$

where $Cost_{cons.\ per\ work\ h}$ is the consumable cost per work hour, \$/work h.

The operating cost per work hour (in \$) can, now, be calculated by:

$$Cost_{operating\ per\ work\ h} = Cost_{cons.\ per\ work\ h} + Cost_{L\ per\ work\ h} \qquad (15.12)$$

where $Cost_{L\ per\ work\ h}$ is the labor cost or wage rate per work hour, \$ (see Example 15.9).

15.7 CALCULATIONS – *WORKED EXAMPLES ON THERMAL ENERGY MACHINING*

Example 15.1: Calculating the Spark Energy in EDM

An RC-type generator is used to machine a metallic workpiece by EDM. The maximum charging voltage applied to the capacitor is 85 volts. The capacitance of the charging capacitor is 110 μF. Calculate the spark energy in the EDM process.

Solution

$C = 110 \ \mu F = 110 \times 10^{-6} \ F$, $V = 85$ volts, $E_s = ?$

By using Equation (15.1),

$$E_s = \frac{CV^2}{2} = \frac{110 \times 10^{-6} \times 85^2}{2} = 0.4 \ J$$

The spark energy = 0.4 J.

EXAMPLE 15.2: CALCULATING THE *MRR* IN EDM

A 10-mm-diameter hole is being produced in a steel workpiece (melting temperature = 1450°C) by using a 90 A current in an EDM process. The depth of the hole is 18 mm. Calculate the *MRR* in the EDM process.

Solution
$I = 90$ A, $T_w = 1450$°C, $MRR = $?
By using Equation 15.2,

$$MRR = 40,000\,I\,T_w^{-1.23} = 40,000 \times 90 \times 1450^{-1.23} = 465.4\,\text{mm}^3/\text{min}$$

EXAMPLE 15.3: CALCULATING THE MACHINING TIME IN EDM

By using the data in Example 15.2, calculate the machining time in the EDM process.

Solution
$MRR = 465.4$ mm³/min, Dia. $= D = 10$ mm, Hole depth $= L = 18$ mm

$$\text{Volume of material removed} = \text{Volume of hole} = \frac{\pi}{4}D^2L$$

$$= \frac{\pi}{4}10^2 \times 18 = 1413.9\,\text{mm}^3$$

$$MRR = \frac{\text{Volume of material removed}}{\text{time taken}} = 465.4\,\text{mm}^3/\text{min}$$

$$\frac{1413.9}{\text{time taken}} = 465.4\,\text{mm}^3/\text{min}$$

$$\text{Machining time} = \frac{1413.9}{465.4} = 3\,\text{min}$$

EXAMPLE 15.4: CALCULATING THE KERF (SLOT WIDTH) IN EDWC PROCESS

An EDWC process is being performed by using a 0.25-mm-diameter wire to machine a 50-mm-thick steel workpiece. The spark gap is 0.05 mm, and the wire feed rate is 100 mm/min. Calculate the kerf (slot width).

Solution
$d_w = 0.25$ mm, s $= 0.05$ mm, $b = $?
By using Equation 15.3,

The kerf or slot width $= b = d_w + 2s = 0.25 + (2 \times 0.05) = 0.25 + 0.10 = 0.35\,\text{mm}$

EXAMPLE 15.5: CALCULATING THE MRR IN EDWC

By using the data in Example 15.4, calculate the *MRR* in the EDWC process.

Solution
$b = 0.35$ mm, $f_r = 100$ mm/min, h = 50 mm, *MRR* = ?
   By using Equation 15.4,

$$\text{MRR} = f_r \, h \; b = 100 \times 50 \times 0.35 = 1750 \, \text{mm}^3/\text{min}.$$

The MRR = 1750 mm³/min.

EXAMPLE 15.6: CALCULATING THE VELOCITY OF ELECTRON IN THE ELECTRON BEAM IN EBM

The voltage between cathode and anode of an electron gun, in an EBM process, is 80,000 V. Calculate the velocity of electron in the electron beam.

Solution
$V = 80,000$ volts; $m = 9.1 \times 10^{-31}$ kg $e = 1.6 \times 10^{-19}$ C
   By using Equation 15.7,

$$\tfrac{1}{2} mv^2 = eV$$

$$(0.5)(9.1 \times 10^{-31}) v^2 = 1.6 \times 10^{-19} \times 80,000$$

$$4.55 v^2 = 12.8 \times 10^{-15}$$

$$v = 0.53 \times 10^{-7} = 5.3 \times 10^{-8} \, \text{m/s}$$

The velocity of electrons = 5.3 × 10⁻⁸ m/s

EXAMPLE 15.7: CALCULATING THE DEPTH OF PENETRATION IN EBM

An electron beam with an accelerating voltage of 150 kV strikes on a steel plate in an EBM process. Calculate the depth of penetration of the electron beam in the plate. Take density of steel as 8000 kg/m³.

Solution
$V = 150$ kV = 150,000 V, Density of steel = $\rho = 8000$ kg/m³ = 8 × 10⁻⁶ kg/mm³
   By using Equation 15.8,

$$d = 2.6 \times 10^{-17} \left(\frac{V^2}{\rho} \right) = 2.6 \times 10^{-17} \left(\frac{150,000^2}{8 \times 10^{-6}} \right) = 73.12 \times 10^{-3} \, \text{mm} = 73.12 \times 10^{-6} \, \text{m}$$

The depth of penetration = 73.12 μm.

EXAMPLE 15.8: CALCULATING THE MRR AND THE CUTTING SPEED IN EBM

A 250-μm-wide slot is cut in a 1-mm-thick titanium plate using a 1.2-kW electron beam during EBM. Calculate the (a) MRR and (b) cutting speed in the EBM process. Hint: take the average value of the unit power range(s) given in Table 3.1.

Solution
Area = 250 × 10^{-3} mm × 1 mm = 0.25 mm², P = 1.2 × 1000 = 1200 W

a. By reference to Table 3.1, P_u = 3.6 W.s/mm³;
   By using the modified form of Equation 3.9,

$$\text{MRR} = \frac{P}{P_u} = \frac{1200}{3.6} = 333.33\,\text{mm}^3\text{/s}$$

b. By using Equation 15.9,

$$v = \frac{\text{MRR}}{A} = \frac{333.33}{0.25} = 1333\,\text{mm/s} = 1.33\,\text{m/s}$$

The cutting speed = 1.33 m/s

EXAMPLE 15.9: CALCULATING THE OPERATING COST PER WORK HOUR FOR A PAC SYSTEM

The nozzle cost and the electrode cost in a PAC system (equipment) are $ 4 and $ 8, respectively. The nozzle/electrode life is 2 arc h. The labor wage rate is $ 30 per work hour, and the duty cycle is 50%. Calculate the operating cost per work hour for the PAC system.
C_n = $ 4, C_e = $ 8, $T_{L(cons.)}$ = 2 arc h, Duty cycle = 50%, $Cost_{L\,per\,h}$ = $ 30
By using Equation 15.10,

$$Cost_{cons.\,per\,arc\,h} = \frac{C_n + C_e}{T_{L(cons.)}} = \frac{4+8}{2} = \$6$$

By using Equation 15.11,

$$Cost_{cons.\,per\,work\,h} = Cost_{cons.\,per\,arc\,h} * \text{duty}(\%) = 6 \times 0.50 = \$3$$

By using Equation 15.12,

$$Cost_{operating\,per\,work\,h} = Cost_{cons.\,per\,work\,h} + Cost_{L\,per\,work\,h} = \$3 + \$30 = \$33$$

The operating cost for the PAC system = $ 33 per work hour.

QUESTIONS AND PROBLEMS

15.1   a. Diagrammatically illustrate the EDM process.
        b. Why is it important to flush new dielectric fluid in and to feed the tool electrode down after each discharge in EDM?
        c. List the applications of EDM.

15.2   Briefly explain the EDWC process with the aid of a diagram.

15.3   a. Diagrammatically illustrate the EBM process.
        b. What are the capabilities/advantages of EBM?

15.4   Briefly explain the LBM process with the aid of a diagram.

15.5   a. What is meant by "plasma"? How is plasma formed in PAC?
        b. Briefly explain the PAC process with the aid of a diagram.
        c. What are the advantages and limitations on PAC?

P15.6   The voltage between cathode and anode of an electron gun, in an EBM process, is 90 kV. Calculate the velocity of electron in the electron beam.

P15.7   An electron beam with an accelerating voltage of 120 kV strikes on an aluminum alloy plate in an EBM process. Calculate the depth of penetration of the electron beam in the plate. Take density of aluminum alloy as 2700 kg/m^3.

P15.8   A 9-mm-diameter hole is being produced in a copper workpiece (melting temperature = 1080°C) by using an 85 A current in an EDM process. The depth of the hole is 20 mm. Calculate the machining time for the process.

P15.9   An RC-type generator is used to machine a metallic workpiece by EDM. The maximum charging voltage applied to the capacitor is 100 V. The capacitance of the charging capacitor is 120 µF. Calculate the spark energy in the EDM process.

P15.10 The nozzle cost and the electrode cost in a PAC system (equipment) are $ 5 and $ 10, respectively. The nozzle/electrode life is 1.7 arc h. The labor wage rate is $ 28 per work hour, and the duty cycle is 40%. Calculate the operating cost per work hour.

P15.11 An EDWC process is being performed by using a 0.25-mm-diameter wire to machine a 40-mm-thick steel workpiece. The spark gap is 0.07 mm, and the wire feed rate is 135 mm/min. Calculate the MRR in the EDWC process.

P15.12 A 300-µm-wide slot is cut in a 2-mm-thick aluminum plate using a 1-kW electron beam during EBM. Calculate the (a) MRR and (b) cutting speed in the EBM process. Hint: take the average value of the unit power range(s) given in Table 3.1.

MCQS

P15.13a. In which NTM process does the temperature exceed 15,000°C?
         (i) EDM, (ii) EDWC, (iii) EBM, (iv) LBM, (v) PAC
    b. Which NTM process can drill 0.1-mm-diameter hole with high MRR?
         (i) EDM, (ii) EDWC, (iii) EBM, (iv) LBM, (v) PAC
    c. Which NTM process is suitable to manufacture metallic tools, molds, and dies?
         (i) EDM, (ii) EDWC, (iii) EBM, (iv) LBM, (v) PAC

    d. Which NTM process equipment requires a mirror and a lens?
        (i) EDM, (ii) EDWC, (iii) EBM, (iv) LBM, (v) PAC
    e. Which thermal energy process is suitable to make long cuts with small kerf?
        (i) EDM, (ii) EDWC, (iii) EBM, (iv) LBM, (v) PAC
    f. Which NTM process causes the most metallurgical damage to the work material?
        (i) EDM, (ii) EDWC, (iii) EBM, (iv) LBM, (v) PAC

REFERENCES

Dahotre, N.B., Harimkar, S.P. (2008), *Laser Fabrication and Machining of Materials*, Springer, Berlin.

Jahan, M.P. (2015), *Electric Discharge Machining*, Nova Science Publishers Inc., New York.

Kumar, K., Kumari, N., Devim, J.P. (2019), *Non-Conventional Machining in Modern Manufacturing Systems*, IGI Global, Hershey, PA, USA.

Okada, A. (2019), *Electron Beam Machining*, Springer, Berlin, Germany.

Owhal, A., Rao, N.S., Gupta, U., Mahajan, M. (2020), Extension of wire-EDM capability for turning titanium alloy and an experimental study for process optimization by grey relational analysis, *Materials Today: Proceedings* 24: 966–974.

Schneider, R.W., (1989), *Electron Beam Machining, In: (ed: ASM), Machining* (Vol. 16), ASM Handbook, Ohio, USA.

16 Electrochemical Machining and Chemical Machining Processes

16.1 ELECTROCHEMICAL MACHINING (ECM) PROCESSES

Electrochemical machining (ECM) processes involve the removal of material by anodic metal dissolution. In ECM processes, electrical energy is used in combination with chemical reactions to accomplish material removal. There are three types of ECM processes: (1) ECM, (2) electrochemical deburring (*ECD*), and (3) electrochemical grinding (*ECG*). These processes are explained in the following sections.

16.2 ELECTROCHEMICAL MACHINING

16.2.1 THE ECM PROCESS AND ITS ADVANTAGES

ECM involves material removal by anodic metal dissolution. In ECM, the metallic workpiece is made the anode by connecting it to the positive (+) pole of a direct current (DC) generator. The tool is made the cathode (−) and is placed in front of the work-area to be machined (see Figure 16.1). The gap between the two electrodes is filled with a conductive electrolytic solution (NaCl or $NaNO_3$), which handles the charge transfer in the working gap (McGeough, 1974; Wilson, 1982).

In ECM, a high amperage DC current is passed between the tool (cathode) and the workpiece (anode) through the electrolyte. The pressurized electrolyte is injected at a set temperature to the area being cut. The ECM tool is advanced (fed) along the desired path close to the work but without touching the piece. The tool feed rate coincides with the rate of "liquefaction" of the material. The gap between the tool and the workpiece varies within 80–800 µm. As electrons (in the electrolytic solution) cross the gap, material from the anodic workpiece is dissolved so the tool forms the desired shape in the workpiece.

The ECM application is limited to metallic/electrically conductive materials. There are certain distinct advantages of ECM; these advantages are listed as follows.

a. The ECM process is capable to machine very hard metals or difficult-to-machine work metals.
b. There is no sparks in ECM process; hence no thermal or mechanical stresses are transferred to the workpiece.
c. High metal removal rates are possible in ECM.

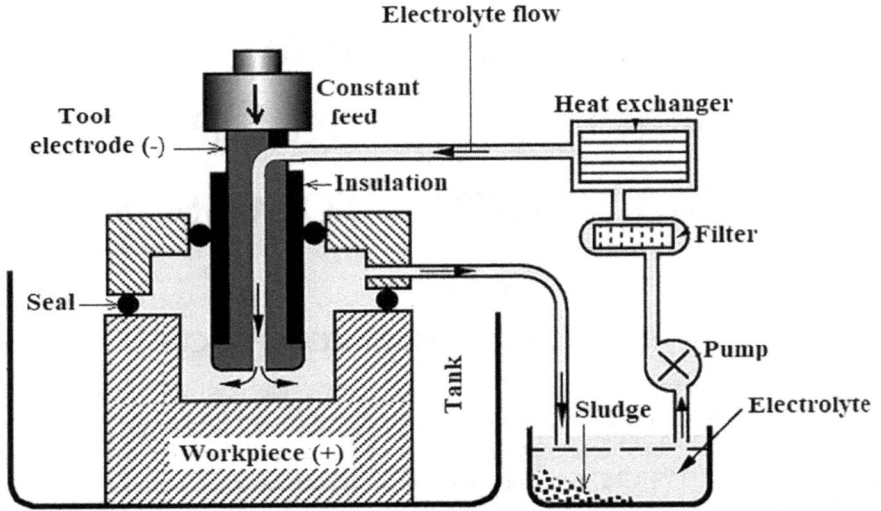

FIGURE 16.1 Electrochemical machining (ECM)

 d. Mirror-like surface finishes can be achieved by machining by ECM.
 e. ECM can cut small or odd-shaped angles, intricate contours or cavities in hard
  and exotic alloys, such as titanium aluminides, superalloys, and the like.

16.2.2 Chemical Reactions in ECM Process

The chemical reactions occurring in the aqueous solution of sodium chloride (NaCl)
during ECM process are presented below.

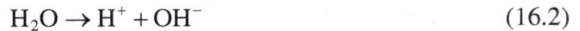

$$NaCl \rightarrow Na^+ + Cl^- \tag{16.1}$$

$$H_2O \rightarrow H^+ + OH^- \tag{16.2}$$

On passing the electric current through the solution, the positive ions move toward
the cathode and negative ions move toward the anode.
 Cathode reactions:

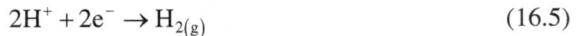

$$Na^+ + e^- \rightarrow Na \tag{16.3}$$

$$Na + H_2O \rightarrow NaOH + H^+ \tag{16.4}$$

$$2H^+ + 2e^- \rightarrow H_{2(g)} \tag{16.5}$$

It is evident in Equation 16.5 that only hydrogen gas will evolve at cathode and there
will be no deposition.

Anode reactions:

$$\text{Fe} \rightarrow \text{Fe}^{2+} + 2e^{-} \qquad (16.6)$$

$$\text{Fe}^{2+} + 2\text{Cl}^{-} \rightarrow \text{FeCl}_2 \qquad (16.7)$$

$$\text{Fe}^{2+} + 2\text{OH}^{-} \rightarrow \text{Fe(OH)}_2 \qquad (16.8)$$

Equation 16.6 indicates that the steel workpiece (anode) is dissolved as Fe^{2+} ions that pass into the aqueous solution thereby forming ferrous chloride ions (see Equation 16.7). The pressurized electrolytic fluid carries away the metal hydroxide formed in the ECM process (see Equation 16.8).

16.2.3 MATHEMATICAL MODELING OF ECM

The ECM is based on the principle of anodic metal dissolution; thus Faraday's laws of electrolysis are applicable. The volume of metal dissolved per unit time (*MRR*) can be calculated by the following mathematical relationship:

$$\text{MRR} = \frac{A\,I}{\rho\,F\,Z} \qquad (16.9)$$

where MRR is material removal rate, cm^3/s; A is the atomic mass of metal (work), g/mol; I is current, ampere (A); ρ is the density of the metal (work), g/cm^3; Z is the valence of the metal; and F = Faraday's constant = 96,500 A-s. In case the MRR is known and the current to be used in ECM is required to be calculated, Equation 16.9 may be re-arranged as follows:

$$I = \frac{\rho * F * Z * (\text{MRR})}{A} \qquad (16.10)$$

The significance of Equation 16.10 is illustrated in Example 16.1.

The rate of dissolution of the work (anode) can be computed by:

$$\frac{dy}{dt} = \frac{\text{MRR}}{\text{Tool area}} \qquad (16.11)$$

where dy/dt is the rate of dissolution, mm/min; and MRR is the material removal rate, mm^3/min (see Example 16.2).

In ECM, the materials to be machined are generally alloys rather than pure metals. It is, therefore, more useful to develop an expression for MRR for alloys, as follows. Let us assume that there are k elements in an alloy. The atomic masses of these elements can be represented as: $A_1, A_2, \ldots, A_k$ with valences $Z_1, Z_2, \ldots, Z_k$. The mass percentages of the elements may be taken as $p_1, p_2, \ldots, p_k$ (in decimal fraction).

The density of the alloy can be calculated by (Huda, 2020):

$$\rho_{alloy} = \frac{1}{\Sigma\left(\dfrac{p_i}{\rho_i}\right)} = \frac{1}{\dfrac{p_1}{\rho_1} + \dfrac{p_2}{\rho_2} + \cdots + \dfrac{p_k}{\rho_k}} \tag{16.12}$$

where ρ_{alloy} is the density of alloy, g/cm³; and $\rho_1, \rho_2, \ldots, \rho_k$ are the densities of the corresponding elements with mass percentages $p_1, p_2,\ldots,p_k$ in the alloy.

On the basis of Equation 16.9, the MRR of the alloy can be calculated by:

$$MRR = \frac{I'}{\left[F\,\rho_{alloy}\,\Sigma\left(\dfrac{p_i Z_i}{A_i}\right)\right]} = \frac{I'}{\left[F\,\rho_{alloy}\left(\dfrac{p_1 Z_1}{A_1} + \dfrac{p_2 Z_2}{A_2} + \cdots + \dfrac{p_k Z_k}{A_k}\right)\right]} \tag{16.13}$$

where MRR is the metal removal rate, cm³/s or mm³/min, and I' is the current per unit valence of the lowest-valence metal in the alloy. The significances of Equations 16.12–16.13 are illustrated in Examples 16.3–16.4.

16.2.4 Machining Parameters in ECM to Achieve a Good Surface Finish

The ECM generates no burr and no stress and provides a long tool life with a damage-free machined surface. It results in a high MRR and good surface quality in engineering alloys, especially those for aerospace applications. In general, it is difficult to obtain good surface quality in ECM of aerospace titanium alloys. Recently, Xuezhen and co-researchers have investigated ECM machining of the aerospace alloy: Ti60 alloy (Ti–5.6Al–4.8Sn–2Zr–1Mo–0.35Si–0.7N) (Xuezhen et al., 2016). They conducted a range of experiments ensuring the stability of the ECM machining process parameters (such as electrolyte temperature, electrolyte flow rate, electrolyte concentration, and spindle motion accuracy) and successfully achieved a good surface finish with a low surface roughness, $R_a = 0.912$ µm in the alloy Ti60. A typical set of results from their experimental work is presented in Table 16.1.

TABLE 16.1
ECM Parameters Data for Machining of Alloy Ti60

#	Machining Parameter	Value
1	Electrolyte concentration	13 wt% *NaCl*
2	Electrolyte temperature	23°C
3	Pulse power voltage	20 V
4	Pulse power frequency	0.4 kHz
5	Pulse duty cycle	0.3 or 30%
6	Anode feed rate	0.5 mm/min
7	Surface roughness, R_a	0.912 µm

16.3 ELECTROCHEMICAL DEBURRING (ECD)

The term "deburring" refers to the removal of burrs or sharp corners on holes in metallic workpieces. *Electrochemical deburring (ECD)* is a slight modification of ECM. It involves the removal of burrs or round sharp corners on holes in metal parts (produced by conventional through-hole by drilling) using the basic principle of ECM.

The *ECD* finishes the workpiece surfaces by means of anodic metal dissolution. In the ECD process, the deburring tool is the cathode (−) that acts under DC current and in the presence of an electrolyte fluid to create the anodic reaction that dissolves the workpiece (+) surface material in a precise manner. The tool areas (close to the work) (where no machining is required) are insulated. In *ECD*, the electrolyte solution transfers charge in the gap between the cathode and the work, which causes electron transfer from the work to remove surface material. The gap between the cathode and the work is key to regulating the material removal process. The shape of the cathode determines the final shape of the work or the impression (imaging) placed upon the work. The speed of material removal is dictated by the DC current applied. The material removed during the deburring process must be filtered out of the electrolyte stream in order to maintain constant electrolyte quality in the gap between the cathode and the workpiece.

16.4 ELECTROCHEMICAL GRINDING (ECG)

Electrochemical grinding (ECG) is a special form of ECM in which grinding wheel with a conductive bond material is used to make more effective the anodic dissolution of metal part surface. The abrasive used in ECG is either aluminum oxide (Al_2O_3) or diamond. In case the bond material is metallic, the abrasive used is diamond, while in case the bond material is resin impregnated with metal particles, the abrasive is aluminum oxide. The abrasive grits jetting out from the grinding wheel at the contact with the workpiece establish the gap distance in ECG (see Figure 16.2).

The ECG has distinct advantage of longer service lives of grinding wheel than conventional grinding. Additionally, the dressing of grinding wheel, in ECG, is less

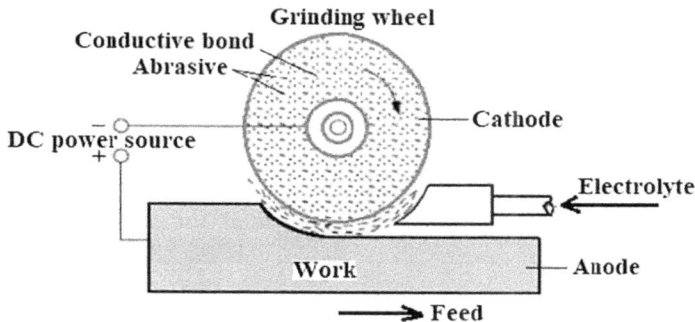

FIGURE 16.2   Electrochemical grinding (ECG)

frequently required. The ECG applications include sharpening of hard cemented carbide tools, grinding of surgical needles, thin wall tubes, fragile parts, and the like. In particular, the present need of processing super-alloys and hard materials with high surface finish as well as with stress and crack-free surface required by industries is demanding for employment of ECG machine (Bhuyan et al., 2020).

16.5 CHEMICAL MACHINING

Chemical machining (ChM) involves the use of a strong acidic or alkaline chemical reagent for controlled chemical dissolution of the work material. In order to protect the areas from which the metal is not required to be removed, special coatings, called *maskants*, are used. The *ChM* finds wide applications in the production of micro-components of micro-electromechanical system (*MEMS*) and semiconductor industry (Bellows, 1982).

The essential steps in ChM include (I) workpiece cleaning, (II) coating clean workpiece with *maskant*, (III) scribing of the *maskant*, (IV) etching (machining process), (V) the workpiece removal, (VI) workpiece rinsing/drying, and (VII) removal of the *maskant* from the workpiece. Once a metallic workpiece has been properly cleaned, it is coated with a *maskant* – a polymeric material (e.g. PVC). The *maskant* must well adhere to the workpiece. The scribing of the *maskant* is guided by templates to expose the areas that receive chemical machining. In the etching (machining process) step, the workpiece is immersed in the selected etchant, and the uncovered areas are machined (see Figure 16.3). Table 16.2 lists the commonly used etchants [*e.g.* hydrofluoric acid (HF) solution, nitric acid (HNO_3) solution, hydrochloric acid (HCl) solution, etc.] for various workpiece metals/alloys.

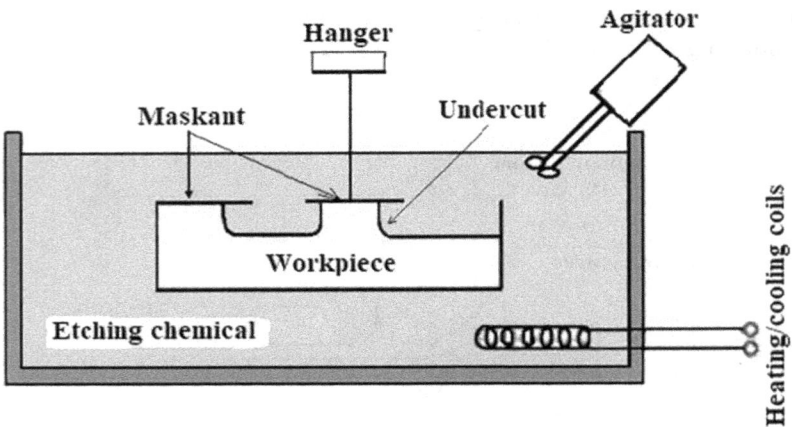

FIGURE 16.3   Chemical machining (ChM)

TABLE 16.2
Metallographic Etchants and Their Proportions for various workpiece metals

#	Workpiece Metal	Etchant	Ratio
1.	Aluminum and alloys	H_2O/HF	1:1
2.	Copper and alloys	H_2O/HNO_3	1:5
3.	Iron/steel	H_2O/HCl	1:1
		H_2O/HNO_3	1:1
4.	Magnesium/alloys	$H_2O/NaOH$ (hot)	10:1 (by wt.)
5.	Nickel and alloys	HNO_3/Acetic acid/Acetone	1:1:1
6.	Titanium and alloys	$H_2O/HF/HNO_3$	50:1:1

16.6 CALCULATIONS – WORKED EXAMPLES ON ELECTROCHEMICAL MACHINING

EXAMPLE 16.1: CALCULATING THE CURRENT REQUIREMENT IN ECM

A nickel plate is being machined by ECM with a required MRR of 570 mm³/min. Calculate the amperage that must be used in the ECM process.

Solution

$$MRR = 570\,\text{mm}^3/\text{min} = \frac{570 \times (0.1\,\text{cm})^3}{60\,s} = 0.0095\,\text{cm}^3/s$$

$A_{Ni} = 58.7\,\text{g/mol}, Z_{Ni} = 2, F = 96,500\,\text{Coulomb}, \rho = 8.9\,\text{g/cm}^3$

By using Equation 16.10,

$$I = \frac{\rho * F * Z * (MRR)}{A} = \frac{0.0095 \times 8.9 \times 96500 \times 2}{58.7} = 278\,\text{A}$$

The required current = 278 A

EXAMPLE 16.2: CALCULATING THE MRR AND THE RATE OF DISSOLUTION IN ECM

In an ECM process of machining pure titanium, a 700-A current is passed through the electrolyte. The tool area is 1000 mm². Calculate the (a) MRR and (b) rate of dissolution.

Solution
$I = 700\,\text{A}, A_{Ti} = 47.9\,\text{g/mol}, \rho_{Ti} = 4.51\,\text{g/cm}^3, Z_{Ti} = 4, F = 96500\,\text{A-s}$, Tool area = 1000 mm²

a. By using Equation 16.9,

$$MRR = \frac{AI}{\rho F Z} = \frac{47.9 \times 700}{4.51 \times 96,500 \times 4} = 0.0192 \, cm^3/s$$

$$= \frac{0.0192 \, (10 \, mm)^3}{\frac{1}{60} \, min} = 1152 \, mm^3/min$$

Metal removal rate = MRR = 1152 mm³/min

b. By using Equation 16.11,

$$\frac{dy}{dt} = \frac{MRR}{Tool \ area} = \frac{1541}{1000} = 1.54 \, mm/min$$

Rate of dissolution = 1.54 mm/min.

EXAMPLE 16.3: CALCULATING THE MRR IN MACHINING OF STAINLESS STEEL IN ECM

A current of 1700 A is being passed during machining of an 18-8 stainless steel by ECM. Calculate the MRR in the process.

Solution

The 18-8 stainless steel contains 18% Cr and 8% Ni and balance iron (Fe).

$$p_{Ni} = 8\% = 0.08, p_{Cr} = 18\% = 0.18, \quad p_{Fe} = 74\% = 0.74$$

By reference to the Periodic Table of Elements and the Physical Properties data, we obtain:

$$A_{Ni} = 58.71 \, g/mol, \rho_{Ni} = 8.9 \, g/cm^3, Z_{Ni} = 2; A_{Cr} = 51.99, \rho_{Cr} = 7.19 \, g/cm^3,$$
$$Z_{Cr} = 2; A_{Fe} = 55.85 \, g/mol, \rho_{Fe} = 7.86 \, g/cm^3, Z_{Fe} = 2$$

$$I' = \frac{Current}{Valence \ of \ the \ lowest \ valence \ metal} = \frac{1700}{2} = 850 \, A$$

By using Equation 16.12,

$$\rho_{alloy} = \frac{1}{\dfrac{p_1}{\rho_1} + \dfrac{p_2}{\rho_2} + \cdots + \dfrac{p_k}{\rho_k}} = \frac{1}{\dfrac{p_{Ni}}{\rho_{Ni}} + \dfrac{p_{Cr}}{\rho_{Cr}} + \dfrac{p_{Fe}}{\rho_{Fe}}} = \frac{1}{\dfrac{0.08}{8.9} + \dfrac{0.18}{7.19} + \dfrac{0.74}{7.86}} = 7.81 \, g/cm^3$$

By using Equation 16.13,

$$MRR = \cfrac{I'}{\left[F\,\rho_{alloy}\left(\dfrac{p_1 Z_1}{A_1} + \dfrac{p_2 Z_2}{A_2} + \cdots + \dfrac{p_k Z_k}{A_k}\right)\right]}$$

$$= \cfrac{850}{\left[96500 \times 7.81\left(\dfrac{p_{Ni} Z_{Ni}}{A_{Ni}} + \dfrac{p_{Cr} Z_{Cr}}{A_{Cr}} + \dfrac{p_{Fe} Z_{Fe}}{A_{Fe}}\right)\right]}$$

$$MRR = \cfrac{850}{\left[96500 \times 7.81\left(\dfrac{0.08 \times 2}{58.7} + \dfrac{0.18 \times 2}{51.99} + \dfrac{0.74 \times 2}{55.85}\right)\right]}$$

$$= 0.0312\,\mathrm{cm^3/s} = \cfrac{0.0312\,(10\,\mathrm{mm})^3}{\dfrac{1}{60}\,\mathrm{min}}$$

$$MRR = 0.0312 \times 1000 \times 60 = 1873.3\,\mathrm{mm^3/min}$$

The MRR = 1873.3 mm³/min

EXAMPLE 16.4: CALCULATING THE MRR FOR SUPERALLOY MACHINING IN ECM

A superalloy is machined by ECM by using a current of 1800 A through the cell. The chemical composition of the superalloy is as follows: Ni = 72 wt%, Cr = 17 wt%, Fe = 5 wt%, Al = 1 wt%, and balance is titanium (Ti). Calculate the MRR assuming the current in the lowest valence metal.

Solution
Current = I = 1,800 A, MRR = ?

$$p_{Ni} = 72\% = 0.72, \quad p_{Cr} = 17\% = 0.17, \quad p_{Fe} = 0.05, \quad p_{Al} = 0.01, \quad p_{Ti} = 0.05$$

By reference to the Periodic Table of Elements and the Physical Properties data, we obtain:

A_{Ni} = 58.71 g/mol, ρ_{Ni} = 8.9 g/cm³, Z_{Ni} = 2; A_{Cr} = 51.99, ρ_{Cr} = 7.19 g/cm³, Z_{Cr} = 2; A_{Fe} = 55.85 g/mol, ρ_{Fe} = 7.86 g/cm³, Z_{Fe} = 2; A_{Al} = 27 g/mol, ρ_{Al} = 2.7 g/cm³, Z_{Al} = 3; A_{Ti} = 47.9 g/mol, ρ_{Ti} = 4.51 g/cm³, Z_{Ti} = 4.

$$I' = \cfrac{Current}{Valence\ of\ the\ lowest\ valence\ metal} = \cfrac{1800}{2} = 900\,\mathrm{A}$$

By using Equation 16.12,

$$\rho_{alloy} = \frac{1}{\Sigma\left(\dfrac{p_i}{\rho_i}\right)} = \frac{1}{\dfrac{p_{Ni}}{\rho_{Ni}} + \dfrac{p_{Cr}}{\rho_{Cr}} + \dfrac{p_{Fe}}{\rho_{Fe}} + \dfrac{p_{Al}}{\rho_{Al}} + \dfrac{p_{Ti}}{\rho_{Ti}}} = \frac{1}{\dfrac{0.72}{8.9} + \dfrac{0.17}{7.19} + \dfrac{0.05}{7.86} + \dfrac{0.01}{2.7} + \dfrac{0.05}{4.51}}$$

$$= 7.95\,g/cm^3$$

By using Equation 16.13,

$$MRR = \frac{I'}{\left[F\,\rho_{alloy}\,\Sigma\left(\dfrac{p_i\,Z_i}{A_i}\right)\right]}$$

$$= \frac{900}{\left[96500 \times 7.95\left(\dfrac{p_{Ni}\,Z_{Ni}}{A_{Ni}} + \dfrac{p_{Cr}\,Z_{Cr}}{A_{Cr}} + \dfrac{p_{Fe}\,Z_{Fe}}{A_{Fe}} + \dfrac{p_{Al}\,Z_{Al}}{A_{Al}} + \dfrac{p_{Ti}\,Z_{Ti}}{A_{Ti}}\right)\right]}$$

$$MRR = \frac{900}{\left[96500 \times 7.95\left(\dfrac{0.72 \times 2}{58.71} + \dfrac{0.17 \times 2}{51.99} + \dfrac{0.05 \times 2}{55.85} + \dfrac{0.01 \times 3}{27} + \dfrac{0.05 \times 4}{47.9}\right)\right]}$$

$$MRR = \frac{900}{767783.4\left(0.0245 + 0.0065 + 0.0018 + 0.0011 + 0.0042\right)} = \frac{900}{767783.4 \times 0.038}$$

$$= 0.0308\,cm^3/s = 0.0308 \times 1000 \times 60 = 1850.85\,mm^3/min$$

MRR = 1850.85 mm³/min

QUESTIONS AND PROBLEMS

16.1   a. State the principle of electrochemical machining (ECM).
        b. Draw a sketch of ECM.
        c. What reactions occur in ECM?
        d. Why is an exhaust fan necessary in the ECM equipment?
        e. List the advantages/limitation of ECM.

16.2   Briefly explain electrochemical deburring (ECD) process.

16.3   a. Explain electrochemical grinding (ECG) with the aid of a sketch.
        b. What are the advantages and applications of ECG?

16.4   a. What are the essential steps in chemical machining (ChM)?
        b. Draw a sketch of ChM.
        c. Why is maskant used in ChM?
        d. What are the applications of ChM?

P16.5 A copper plate is being machined by passing a current of 1900 A in ECM. Compute the (a) material removal rate and (b) rate of dissolution if the tool area is 1200 mm²

P16.6 Electrochemical machining of wrought iron is to be performed with material removal rate requirement of 600 mm³/min. What amperage must be used for the ECM process?

P16.7 A superalloy is being machined by ECM by using a current of 1700 A through the cell. The chemical composition of the superalloy is as follows: Ni = 74 wt%, Cr = 15 wt%, Fe = 4.8 wt%, Al = 1.2 wt%, and balance is titanium (Ti). Calculate the metal removal rate, assuming the current in the lowest valence metal.

P16.8 A current of 1650 A is being passed during machining of an 18-2 stainless steel by ECM. Calculate the material removal rate in the process.

REFERENCES

Bellows, G. (1982), *Chemical Machining: Production with Chemistry*, 2nd Edition, Machinability Data Center, Met-cut Research Inc., Cincinnati, OH, USA.

Bhuyan, B.K., Garg, C., Gupta, L. (2020), Design and development of table-top electrochemical grinding set-up, *Materials Today: Proceedings*, 21 Part 32020: 1479–1482.

McGeough, J.A. (1974), *Electrochemical Machining*, Springer, Berlin, Germany.

Huda, Z. (2020), *Metallurgy for Physicists and Engineers*, CRC Press, Boca Raton, FL, USA.

Wilson, J.F. (1982). *Practice and Theory of Electrochemical Machining*, R.E. Krieger Publishing Company Inc., FL, USA.

Xuezhen, C., Zhengyang, X., Dong, Z., Zhongdong, F., Di, Z. (2016), Experimental research on electrochemical machining of titanium alloy Ti60 for a blisk, *Chinese Journal of Aeronautics*, **29**(1), 274–282.

Answers

CHAPTER 2

P2.7.   (a) 0.573, (b) 32.17°.
P2.9.   1.977.
P2.11. 0.742.

CHAPTER 3

P3.5. $F = 1171.2$ N, $N = 1148.2$ N, $F_s = 730$ N, $N_s = 1468.67$ N.
P3.7.

Merchant's equation method	$\dfrac{F}{N} = \tan \beta$ method	Merchant's force circle method
$\beta = 46°$	$\beta = 45.5°$	$\beta = 46.4°$

P3.9.   % specific energy for shear = 40.9 (or the specific energy for shear = 40.9%).

CHAPTER 4

P4.7.   291.4 min.
P4.9.   99.98.
P4.11. $V T_L^{0.1} = 114.55$.

CHAPTER 5

P5.1.   (MCQs). (a) ii, (b) iii, (c) i, (d) iv, (e) i, (f) ii, (g) i, (h) v, (i) i.
P5.7.   895.46 rev/min.
P5.9.   (a) 2, (b) $N_{(1)} = 143.22$ rpm, $N_{(2)} = 147.65$ rpm, $D_{f(1)} = 97$ mm, $D_f = 94$ mm, (c) 20.63 min.
P5.11. 4.8°.
P5.13. (a) 1:15, (b) 8.67 mm.

CHAPTER 6

P6.7.   (a) 40 ft./min = 12,192 mm/min, (b) 0.002 in/rev = 0.0508 mm/rev.
P6.9. $F_c = 1542$ N.
P6.11. 4.8 mm.
P6.13. 7.7 cm.

CHAPTER 7

P7.7.   (a) 0.2 mm/tooth, (b) 144 cm³/min.
P7.9.   (a) 3.8 s, (b) 448.12 cm³/min.

P7.11. (a) up milling, (b) 5.43 s, (c) 1167 cm^3/min.
P7.13. Option 1: for each division, 5 full revolutions and 15 holes on the
21-holes circle of the. Plate No. 2 of the Brown and Sharpe Co. Option
2: for each division, 5 full revolutions and 35 holes on the 49-holes
circle of the Plate No. 3 of the Brown and Sharpe Co.
P7.15. For each division, the index crank must turn 20 holes on the 29-holes
circle of Plate # 2.
The differential change gears are: $Z_1 = 40$, $Z_2 = 29$, $Z_i = 24$, $Z_4 = 48$ teeth;
One idle gear.
7.17.   (a) iii, (b) i, (c) iv.

CHAPTER 8

P8.7. $L_{s(max)} = 233$ mm.
P8.9.   (a) 0.667, (b) 216°.

CHAPTER 9

P9.5.   76.365 kN.
P9.7.   12 mm.

CHAPTER 10

P10.7. 25.08 mm.
P10.9. 0.129 mm.
P10.11. 2.5 mm.
P10.13 (MCQs). (a) iv, (b) i, (c) iii.

CHAPTER 11

P11.9. 3.82 μm.
P11.11. 1.41 mm.
P11.13. 60 × 10^5 chips/min.
P11.15 (MCQs): (a) ii, (b) iii, (c) i, (d) iii.

CHAPTER 12

P12.5. 11.8%.
P12.7. 9 x 10^6 (μm)^3/s.
P12.9. 4.14 min.
12.10. (a) ii, (b) i, (c) iv, (d) iii.

CHAPTER 13

P13.9. N10 G90 G94 G96.
N20 G03 X5.5 Y4.0 R10.0 T02 S2000 F110.

P13.11. 0.75 s.

P13.13. 1750 pulses.

P13.15. (a) Work offset: G54, Tool length compensation: H01.
(b) P: X = 20 Y = 20; Q: X = 20 Y = 180; R: X = 180 Y = 180; S: X = 180 Y = 20.

13.17. MCQS: (a) i (b) iv, (c) ii, (d) iii.

CHAPTER 14

P14.5. 43.68 g.

P14.7. 594.46 mm³/s.

P14.9. 18.61 μm.

14.11 (MCQs): (a) ii, (b) i, (c) ii, (d) iv.

CHAPTER 15

P15.7. 138.67 μm.

P15.9. 0.6 J.

P15.11. 2106 mm³/min.

P15.13 (MCQs): (a) v, (b) iii, (c) i, (d) iv, (e) ii, (f) v.

CHAPTER 16

P16.5. (a) 4186.13 mm³/min, (b) 3.48 mm/min.

P16.7. 1759.35 mm³/min ≅ 1760 mm³/min.

Index

A

Abrasive machining processes, 5, 181–211
Abrasive finish machining operations, 201–211
Abrasive finish machining calculations, 206–211
Abrasive materials, 183
Abrasive jet machining (AJM)
    advantages, 237
    calculations, 244–245
    mathematical modeling, 239
    process, 237
    process parameters, 238
Abrasive water jet machining (AWJM)
    advantages, 241
    calculations, 246–248
    mathematical modeling, 242
    process, 241
Analysis, *see* Engineering analysis
Arithmetic average (R_a), 185
Automatic lathe, 67

B

Boring, 64
Broach
    defined, 148
    design analysis, 149
    materials, 148
    parts, 148
    types, 148
Broaching
    Broaching defined, 147
    Broaching applications, 148
    Broaching analysis/mathematical modeling, 150
    Broaching calculations/worked examples, 151–152

C

Calculations
    in AJM, 244–245
    in AWJM, 246–248
    in broaching, 151–153
    in chip formation mechanics, 19–24
    in CNC machining, 226–231
    in cutting tool lives, 49–56
    in drilling operations, 100–104
    in EBM, 259–260
    in ECM, 269–272
    in EDM, 257–258
    in forces and power in machining, 33–40

    in gear hobbing, 171–174
    in gear milling, 168–169
    in gear shaping, 170
    in milling operations, 119–128
    in shaping operations, 140–143
    in turning operations, 75–86
    in USM, 242–244
    in WJM, 246
Center-less grinding, 188; *see also* Grinding machines
Chemical machining, 268
Chip, 11
Chip formation, 11
Chip formation analysis, 13
Chip formation calculations, 19–24
Chip formation mechanics, 11–26
Chip types
    Continuous chips, 17
    Continuous chips with BUE, 18–19
    Discontinuous chips, 17
    Serrated chips, 18–19
Chip thickness
    Chip thickness ratio, 13
    Uncut chip thickness, 13
    Deformed chip thickness, 13
Clearance angle, 14
Coefficient of friction, 15
Computer aided design (CAD), 217
CAD role in CNC machining, 217
Computer numerical controlled (CNC) machining
    advantages/limitations, 215
    calculations, 226–231
    CNC code words/codes, 219
    CNC lathe, 215
    CNC machine's working system, 217–221
    CNC system, 215
    CNC (work-part) programming, 217–221
    components, 217–221
    milling machine, 221
    positioning systems, 222
    Zero, 221
Control systems in CNC machines
    open-loop control CNC system, 223
    closed-loop control CNC system, 224
Conventional machining processes, 4
Coordinate positioning systems, 222
Cost of machining, 257
Contour turning, 64
Counter boring, *see* Drilling-related operations
Counter sinking, *see* Drilling-related operations
Cutoff (by turning), 64
Cutting force, 27

Cutting force measurement, 28
Cutting forces calculations/worked examples, 33
Cutting fluids, 48
Cutting speed, 69, 96, 113
Cutting time, 69, 96, 114
Cutting power, 69, 115
Cutting tools
    defined, 45
    failure, 46
    life, 47
    materials, 45
    types, 45
    wear, 46
Cylindrical grinding, *see* Grinding machines

D

Decoder, 220
Discontinuous chips, *see* chips
Drill bits
    function, 93
    materials, 93
    types, 93
    twist drill bit, 93
Drilling
    applications, 92
    calculations/worked examples, 100
    defined, 91
Drilling machines
    features, 94
    multiple-spindle drill press, 95
    radial arm drill press, 95
    Turret drill press, 95
    types, 94
    upright drill press, 94
Drilling-related operations
    boring, 92
    counter boring, 92
    counter sinking, 92
    reaming, 92
    tapping, 92
Down milling, *see* Milling methods
Direct indexing, *see* Indexing in Milling
Differential indexing, *see* Indexing in Milling
Design analysis of broach, *see* Broach
Dynamometer, 28

E

End milling, 111
Engineering analysis/mathematical modeling
    of AJM, 239
    of AWJM, 242
    of broaching, 93
    of cylindrical grinding, 187
    of drilling, 96
    of EBM, 254

    of ECM, 265
    of EDM, 252
    of gear milling, 159
    of gear shaping, 163
    of gear hobbing, 166
    of honing, 203
    of lapping, 204
    of milling, 113–115
    of shaping, 136–138
    of surface grinding, 150
    of turning, 69–74
    of USM, 236
    of WJM, 240
Engine lathe, 66; *see also* Lathe machine tools
Electric discharge machining (EDM)
    applications, 251
    mathematical modeling, 252
    process, 251; *see also* Thermal energy
        machining processes
Electron beam machining (EBM)
    mathematical modeling, 254
    process, 254
Electric discharge wire cutting (EDWC)
    mathematical modeling, 253
    process/advantages, 252
Electrochemical machining (ECM)
    advantages, 263
    mathematical modeling, 265
    process, 263
    reactions, 264
Electrochemical de-burring (ECD), 267
Electrochemical grinding (ECG), 267

F

Facing, 64
Face milling, 111
Feed, 3, 69
Feed rate, 113
Finishing/finish machining, 7
Forces in machining, 27
Form turning, 64
Friction force in machining, 27

G

G-codes in CNC programming, 219
Gear (spur gear) nomenclature, 157
Gear cutting/manufacturing methods, 158
Gear cutting calculations, 168–171
Gear broaching, 161
Gear hobbing process, 164
Gear hobbing analysis/mathematical modeling, 166
Gear milling
    analysis/mathematical modeling, 159
    calculations, 168
    process, 159

Gear shaping analysis/mathematical modeling, 163
Gear shaping process, 162
Grinding operation
    analysis/mathematical modeling, 189–191
    calculations, 192
    grinding operation defined, 181
    grinding advantages, 181
    grain cutting action in grinding, 182
Grinding wheel, 182
Grinding wheel parameters, 183
    bonding material, 183
    grade, 183
    Grit number, 183
    structure, 183
Grinders/grinding machines
    center-less grinding machine, 188
    cylindrical grinding machine, 187
    grinding machine features, 187
    surface grinding machine, 187
    types of grinding machines, 187
Grooving, 64

H

Hobbing, 164
Hobbing tool/hob, 164
Horizontal-spindle milling machine, 109
Horizontal-spindle surface grinding machine, 187
Honing operation, 201
Honing applications, 201
Honing mathematical modeling, 203
Honing calculations, 206

I

Industrial importance/applications
    of abrasive finish machining, 201
    of broaching, 147
    of calculations in machining, 8
    of CNC machining, 215
    of cutting forces and power, 27
    of drilling operations, 91
    of ECM, 263
    of machining, 4
    of mechanical energy NTM, 235
    of milling operation, 107
    of shaping/planing operation, 133
    of thermal energy NTM, 251
    of turning operations, 63
Indexing head, 116
Indexing methods, 117
    direct indexing, 117
    plain indexing, 117
    differential indexing, 118
Indexing in milling, 116
Interpolator, 220

K

Knee-and-column type milling machines, 109
Knurling, 64
Kerf, 253

L

Lapping operation, 203
Lapping advantages and applications, 204
Lapping mathematical modeling, 204
Lapping calculations, 206
Laser beam machining (LBM), 255
Lathe machine parts and mechanism, 65
Lathe machine tools
    automatic lathe, 67
    engine lathe, 66
    speed lathe, 66
    tool-room lathe, 66
    turret lathe, 67

M

Machinable materials, 8
Machinability, 7
Machine control unit (MCU), 220
Machine tools
    CNC machines, 215
    drilling machines, 94–95
    gear cutting machines, 165
    grinding machines, 187–188
    lathes, 66–67
    milling machines, 108–111
    shaping/planing machines, 134–135
Machining
    classification of machining, 215
    roughing and finishing in machining, 7
    orthogonal and oblique machining, 11–12
Machining allowance, 3
Machining parameters
    (or cutting condition), 3
Machining processes
    conventional machining processes, 4
    abrasive machining processes, 5
    non-traditional machining (NTM), 5
    micro-precision machining, 6
    ultra-precision machining (UPM), 6
Material removal rate (MRR), 68, 113
Mathematical modeling in machining,
    see Engineering analysis of machining
MCU components
    program reader, 220
    decoder, 220
    interpolator, 220
Mechanical energy NTM processes, 235
Merchant's equation, 16
Merchant's Force Circle, 31

Micro-precision machining, 6
Milling
    analysis/mathematical modeling of
        milling, 108
    calculations in milling, 119
    forms of milling, 107
    indexing in milling, 116
    methods of milling, 108
    milling defined, 107
Milling cutters, 108
Milling machines
    knee-and-column type milling machines, 109
    universal horizontal milling machines, 110
    ram-type milling machines, 110
Milling operations, 111

N

Non-traditional machining (NTM) processes, 5,
    235–273
Normal force, 27
Normal to shear force, 27

O

Orthogonal cutting/machining, 11
Oblique cutting/machining, 12

P

Part programming in CNC machining, 219
Plasma arc cutting (PAC), 255
Planing operation, 133
Planer parts and function, 134
Planer working principle, 134
Pocket milling, 111
Positioning systems in CNC machining, 222
Polishing (abrasive finish machining), 205
Power in machining, 31
Processing equipment in CNC machining, 221
Profile milling, 111

Q

Quick-return mechanism (QRM) in
    shaping, 136
Quality of surface, 185–186

R

Rake angle, 13, 15
Ram-type milling machines, 111
Roughing in machining, 7
Rough surface, 185

S

Serrated chips, 19
Shear force, 15
Shear plane angle, 15
Shear plane area, 15
Shear stress, 15
Specific energy in machining, 31
Speed lathe, 66
Straight turning, 64
Slab milling, 111
Slotting/slot milling, 111
Side milling, 111
Straddle milling, 111
Surface grinding machine, 187
Surface quality, 185
Surface roughness, 185
    measurement, 186
    parameters, 185
    tester, 186
Super-finishing, 205
Super-alloy machining in ECM, 268

T

Taylor's Tool-Life Equation, 47
Temperature in machining, 33
Thermal energy NTM processes, 251
Thrust force, 27, 28
Threading or thread turning, 74
Tool failure, 46
Tool life, 47
Tool-room Lathe, 66
Tool wear, 46
Taper turning mathematical modeling, 72
Turning operation, 63
Turning related operations, 64
Turret Lathe, 67

U

Ultra-precision machining (UPM), 6
Ultra-sonic machining (USM), 235–236
Universal horizontal milling machines, 110

W

Worked examples, *see* Calculations
Work-holding techniques in lathe practice, 68
Wire EDM, *see* Electric discharge wire cutting
    (EDWC)

Z

Zero systems in CNC machining, 221

For Product Safety Concerns and Information please contact our EU
representative GPSR@taylorandfrancis.com
Taylor & Francis Verlag GmbH, Kaufingerstraße 24, 80331 München, Germany